THE IMPACT OF MICROELECTRONICS TECHNOLOGY

The Impact of Microelectronics Technology

EDITED BY MERVYN A. JACK

EDINBURGH
University
Press

22 George Square, Edinburgh

Set in Linoterm Plantin by
Speedspools, Edinburgh
and printed in Great Britain by
Redwood Burn Limited
Trowbridge, Wilts

British Library Cataloguing in Publication Data

Jack, Mervyn A.
The impact of microelectronics technology
1. Microelectronics—Social aspects—Great Britain
I. Title
303.4'83 TK7874

ISBN 0 85224 426 6

CONTENTS

PREFACE

With a dynamic technology such as microelectronics in which technological innovations, new manufacturing techniques, novel circuit concepts and new applications develop almost overnight, it is a severe task for any individual to keep abreast of these developments. It therefore seems appropriate that this text on microelectronics technology has been written by a group of contributors, many of whom are internationally renowned in specific aspects of microelectronics such that each contribution might benefit from the individual specialisation of the contributor.

It is of further significance that the contributors all work under one roof and are all members of the Department of Electrical Engineering at Edinburgh University and its associated Wolfson Microelectronics Institute – few academic establishments in the world can boast a team with such wide-ranging expertise in microelectronics technology. The all-embracing, in-depth treatment of microelectronics presented here will do much to create an impression of the technology, its impact and its implication to pupils and students alike as well as to members of the general public.

J. H. Collins

THE INTEGRATED CIRCUIT

WHAT MICROELECTRONICS IS ALL ABOUT

Microelectronics, one of the most highly developed mass-production technologies, permits realisation of more than half-a-million electronic components on a single chip of silicon some 5 mm square. In this way, highly complex electronic circuits can be packaged in an integrated circuit that measures about 50 mm by 10 mm. The cost-effectiveness and reliability of such integrated circuits have led to their widespread application in all types of microelectronics-based products and microelectronics is now a major wealth-producing industry world-wide.

Photograph courtesy of Wolfson Microelectronics Institute.

INTRODUCTION

Since the early seventies the Department of Electrical Engineering at the University of Edinburgh has specialised in teaching and research in microelectronics technology. As a direct result of this specialisation the Wolfson Microelectronics Institute was formed in association with the academic Department and, over the years, the Department and the Institute have established the University of Edinburgh as a 'centre of excellence' in microelectronics on both a national and an international basis.

The existence of this centre of excellence has meant that increasingly the members have been called upon to lecture and talk on microelectronics not only to other interested academics and to industry at both engineering and managerial levels but also to the general public, and it was as a direct result of these activities that the concept of this book arose.

Each chapter has been contributed by members of the staff of the Department and of the Institute, each specialising in a specific aspect of microelectronics. Chapter 1 surveys the field of microelectronics, highlighting the main differences between microelectronics and conventional electronics. It reviews the history of microelectronics technology from the first electronic computers to the present day, where national initiatives are under way in the search for integrated circuits with up to one million components on a small chip of silicon. Here the problem of designing integrated circuits is contrasted with the problem of customising standard integrated circuits, such as microprocessors, by means of software.

Chapter 2, 'Microelectronic circuit fabrication', considers in detail the various steps involved in the manufacture of integrated circuits. Although processing techniques are continually improving, the basic manufacturing principles have remained unchanged over the years and this chapter describes the basic processing steps with details of mask making, photolithography and ion implantation. Chapter 3 considers the all-important microprocessor, a highly versatile integrated circuit type that is the main reason for the increased public awareness of microelectronics. This chapter attempts to create a perspective of the microprocessor in the overall realm of microelectronics, showing how succeeding generations of microprocessor have developed in technological terms at a far greater rate than they have been applied to products and systems.

From this point forward the basic technological details are assumed and the book turns to applications and product developments based on microelectronics. Chapter 4, entitled 'Home and office computers', provides a useful guide and explanation for the types and varieties of computer currently available for both home and office use. From a discussion of the basic principles on which computers are based, this chapter considers the use of personal computers in the home, word processors, microcomputers and minicomputers in the office and large mainframe computers in terms of their use for large database storage and retrieval.

Home computers are only one aspect of the new range of microelectronics-based products available to the consumer, and in chapter 5 several new product areas are reviewed to identify trends in the impact of microelectronics on consumer products. The importance of the television set is emphasised as a centre for home entertainment and information. Facilities based on the home computer and developments in viewdata and teletext are considered in detail, as are video cassettes and video discs. The application of microelectronics to toys and games is reviewed in terms of 'speaking' products and television games, and the emerging use of microelectronics in the motor car is discussed.

Another aspect of microelectronics that affects the general public is the development of new medical products, and chapter 6 reviews some of the more important developments in this field. From the invaluable use of microelectronics in medical diagnostic equipment such as X-ray, ultrasonic scanner and ECG systems, the chapter extends to the use of microelectronics in prescription labelling and the emerging use of microelectronics by the general practitioner for patient records and diagnosis.

Attention is transferred to the industrial scene in chapter 7, which is entitled 'Microprocessors in industrial process control'. Here the application of microelectronics to provide 'intelligence' and automation in complex industrial processes is highlighted, and the use of condition-monitoring equipment for improved reliability, safety and efficiency is discussed.

Chapter 8 addresses the all-important aspect of the reliability of microelectronics and products based on microelectronics. From a consideration of the various failure modes the chapter goes on to discuss techniques that can be applied to enhance product reliability and to examine the reliability of a product over its design life. The reliability of microelectronics devices themselves is highlighted and testing methods for these devices are described. The final chapter draws together much of the material in the book and discusses some of the social impact of microelectronics technology. Use of microprocessors in the home, in schools and in the office is discussed and the need for society to accept the challenge of microelectronics to create increased wealth is identified.

Microelectronics, although not guaranteed to be the technology of the future, is certainly a technology with a future, a future that will affect society in general and each individual in it. This text represents an attempt by the technologists to make society more aware of the technology, its implications, its potential, its limitations and its value. We hope it will help to engender optimism for the future of a society pervaded by microelectronics.

M. A. Jack

1. MICROELECTRONICS : THE SCENE TODAY

Microelectronics is the technology that makes it possible for a tiny chip of silicon to incorporate, with high reliability, the functions that previously required thousands of individual transistors (and, before them, valves). Thus microelectronics, or integrated electronics, differs from the established electronics of the past forty years because the prime emphasis is now on the silicon integrated circuit or chip. All traditional components, such as transistors or wires, can be interconnected in solid form within the silicon itself and, moreover, most other components can themselves be formed from transistors.

The aim of all electronic and microelectronic engineering is to produce equipment or products that can perform their functions in the smallest size, with the lightest weight and at the lowest cost. The prime emphasis behind all electronics developments is to shrink the constituent parts – to miniaturise the hardware – to miniature or microminiature form. This has led to electronic equipment that is smaller, lighter, lower in power, less costly and more reliable than equipment with assemblies of separate components. Integrated circuit electronics has been used in every type of electronic system and has revolutionised many fields, including industrial electronics, medical instrumentation and consumer electronics. Representative products in the latter category are the ubiquitous pocket calculator and the transistor radio. With both these products, the electronic system has been reduced to the input/output transducers – such as the keyboard, speakers and displays, which convert physical quantities into electrical quantities and vice versa – with the electronic intelligence being provided by the silicon chip. Expensive industrial assembly operations have been minimised, with the real design and engineering being contained within the integrated circuits. Thus the reliability of microelectronics-based products is extremely high, with overall system reliability being more dependent upon the properties of the packaging and its connections than on the reliability of the integrated circuit itself.

The actual differences between microelectronics and traditional electronics are striking enough for microelectronics to be regarded as a new industry. It is so intensely scientific that new companies will not thrive and grow unless highly qualified manpower can be attracted to them. Coupled with this labour requirement is the need for intensive capital investment in order that the required microcircuits can be designed, fabricated and tested efficiently. Indeed, a modern microelectronics manufacturing plant typically requires an initial capital investment of over $20m, excluding the costs of a purpose-built plant and its services. However, in spite of the high investment levels required, the main challenge for management lies in maintaining the necessary number of highly qualified personnel, from design and applications specialists to skilled manpower for processing and testing the microcircuits. The subject of microelectronics is thus multi-disciplinary, with the skills required ranging from solid-state physics through electronic engineering, physics and chemistry to mechanical engineering. It is rare to find individuals who have expertise in more than one of these areas. However, to optimise the continuing development of the technology, it is vital that the engineers and scientists employed in the microelectronics industry have an appreciation of its many facets.

The subject of electronics – as opposed to microelectronics – goes back to the early years of the twentieth century with the invention of the vacuum tube or electronic valve. This was a development of the simple electric light bulb which, instead of just producing light, allowed electrical signals – composed of electrons – to be amplified. This invention led to the development in the twenties and thirties of both radio and television and produced the first widespread impact of the technology on society.

About this time the first electronic computers were being produced, making use of the same electronic valves that were used in radios. It is difficult to appreciate the scale of these early computers because they bear little resemblance to modern computers. One of the first truly electronic computers was unveiled in 1940, at the University of Pennsylvania. Called the ENIAC, it filled a room ten metres by fifteen metres, weighed thirty tonnes, and was built with 18,000 individual valves. Its power consumption was equivalent to 180 single-bar electric fires and, in spite of its size, it was only capable of carrying out the simple calculations that are now done by millions of people on their pocket calculators, which weigh only a few ounces, are battery powered and cost a few dollars.

It is the development of the computer and its microelectronic equivalent, the microprocessor, as opposed to improvements through the use of microelectronics in radio and television, that is currently providing the greatest challenge to society. The ready availability of computing power is allowing many new products to be developed. How then is this possible, and how has it come about? The key to both these questions is the transistor, which was first demonstrated in 1947 at the Bell Telephone Laboratory near New York. The advantages of the transistor over the older valves were that it was very much smaller in size, used much less electrical power and subsequently proved to be very much more reliable. The invention of the transistor was the breakthrough that provided the impetus to the electronics industry, and during the period from 1947 to 1958 the technology of semiconductor transistors was perfected, giving rise to a new industry now known as the semiconductor industry. By 1959, there were fifty-seven semiconductor companies in the United States alone, operating in a very labour-intensive industry.

The invention of the bipolar transistor, although enabling engineers to produce electronic equipment that was smaller, lighter, more reliable, of higher performance and less costly than equipment based on thermionic valves, was only the prelude to a much greater revolution – that of the monolithic integrated circuit. This development enabled all the individual components of an electronic circuit to be merged into a solid or integrated form. Early integrated circuits were very simple and had only a few tens of

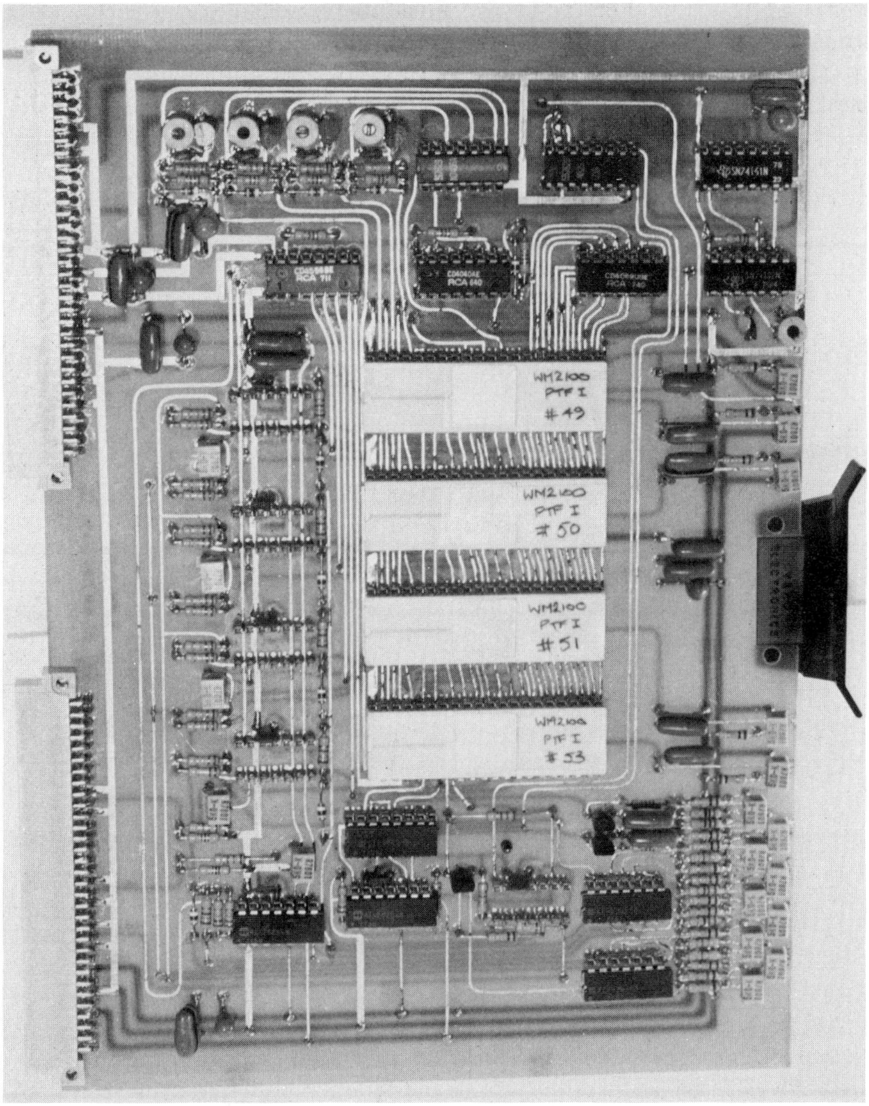

Electronic circuit card. A modern printed-circuit card holds many individual microelectronics devices. Normally the card will contain several LSI microcircuits (large white packages) with a number of peripheral MSI chips (black packages) and a collection of discrete components. The LSI microcircuits shown here each contain a total of 5000 transistors, and the card forms part of an 'intelligent' electronic instrument.

Photograph courtesy of Wolfson Microelectronics Institute.

components per chip. This level of integration, achieved during the early sixties, was referred to as small-scale integration (or SSI). By the mid-sixties, the level of integration brought about by improvements in the semiconductor industry had risen to several hundred components, corresponding to medium-scale integration (MSI). The seventies were the age of large-scale integration (LSI), where integration levels ranged from several thousand to a quarter of a million components per chip. Logically, therefore, the eighties are already being referred to as the VLSI era, where VERY large-scale integrated circuits will be produced in volume.

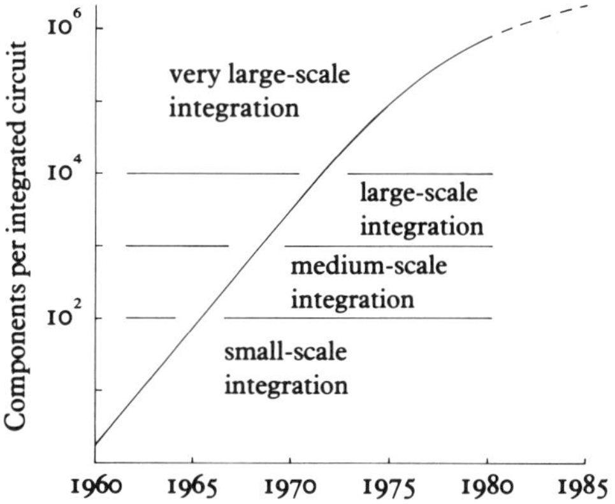

The changing scale of integration. The dramatic increase in the number of components that can be fabricated on each integrated circuit has developed from small-scale integration (SSI) in the early sixties, through medium-scale integration (MSI) and large-scale integration (LSI) to the present levels of very large-scale integration (VLSI) with up to one million components per circuit. This growth (known as 'Moore's Law' after Dr G. Moore of Intel who first predicted it) represents a doubling of circuit complexity approximately every two years and is due, in part, to the availability of larger circuits coupled with improvements in circuit design methods. However, the main feature governing this growth has been the ability to make smaller and smaller devices, thereby allowing more devices to be fabricated in the same silicon area.

One of the father figures of the microelectronics industry, Gordon Moore – formerly of Fairchild Semiconductors and a founder of the Intel Corporation – predicted in the early sixties that the number of components that would be integrated into a single integrated circuit would double every year for the then foreseeable future. Remarkably, this has been achieved, and it looks as if what has become known as Moore's Law will be obeyed for several years to come. From a few components in the early sixties this progression has lead to over 100,000 transistors in a single complex silicon chip today. Developments still continue, and the prospect of having many millions of integrated components in the near future is actively being pursued. What is perhaps more significant is that whereas the initial research and development of transistors, together with the appropriate production technology, was undertaken in the USA, it now looks certain that Japanese engineers will also contribute significantly in the next decade. This is anticipated largely because nationally orchestrated programmes already exist within Japan, and the Japanese are dedicating enormous funds to centralised VLSI research. Because of the strategic importance of the semiconductor industry in both military and commercial terms it is likely that other countries will attempt to respond to the Japanese challenge and try to dominate in some aspects of the microelectronics field.

The key reason why integrated circuits or chips are all-pervasive in a variety of applications is their very low cost. These complex components are made very cheaply through the application of mass-production techniques, and microelectronics technology is an outstanding example of automation and efficiency. The raw materials are silicon, aluminium, water and a variety of common gases, all of which ensure a low cost product. The silicon-processing industry has now been established for more than twenty years, and standard silicon processes are widely available throughout the world. Modern processing is based on so-called planar technology, a significant landmark introduced in 1959, permitting the formation of the monolithic silicon integrated circuit with electrical connections accessed on only one face of the device. Earlier devices required contacts to both sides of the semiconductor, a practice that was difficult to introduce into a production environment. The advent of the planar process also permitted each of the individual components contained within the chip to be individually isolated from one another by means of 'junctions' and it allowed each component to be defined using a patterning process, based on the important properties of the natural oxide of silicon, a glass-like insulator known as silicon dioxide or silica.

The process of making integrated circuits is complex in its detail, but is conceptually easy to appreciate in essence. Any silicon fabrication process involves a large number of individual processing operations performed on the basic silicon starting material. The silicon is produced from special sand, from which the oxygen is removed, to leave a perfect silicon crystal with very few impurities. The silicon is refined in special furnaces to produce a large solid form, typically 50–150 mm in diameter and up to 2 m long, known as a 'boule'. The boule is sliced into individual wafers, and it is on these wafers that hundreds of integrated circuits can be made at one time. Each transistor is made from a number of junctions, which are themselves formed by deliberately introducing into the basic, pure silicon material small amounts of other materials called, ironically, 'impurities'. The size and positions on the surface of the wafer of each of the necessary component junctions are defined by a photographic printing process. Essentially, where junctions are to be formed, holes are opened up in the protective silicon dioxide barrier such that, when the wafers are subsequently placed in an impurity gas environment at high temperature, junctions will be formed where the bare silicon has been exposed through the holes. The difficult part of this photolithographic process is that, in order to produce the many thousands of transistors in the space of 5 mm square, the dimensions of the individual components have to be very, very small. In the latest microprocessors, for instance, which are particular types of VLSI, the smallest element of the circuit will be about two microns or two millionths of a metre (2 μm) in length. Nothing in everyday life is that small: the nearest thing is a human hair, which is about fifty times as big. Anyone familiar with photography will be acquainted with the problem of keeping dust particles off negatives, but consider how much more difficult it is when almost invisible particles of size less than half a micron have to be excluded from a whole manufacturing area. It requires the machine operators in the semiconductor factories to wear not only special coats but hoods and boots also, producing a spaceman-like appearance. While these very small dimensions are essential to permit hundreds of thousands of transistors to be made in one integrated circuit about 5 mm square, the circuits themselves are mass produced, and many hundreds are fabricated in a thin wafer of silicon material which is subsequently cut up to form each individual circuit. The wafers are processed in batches of about one hundred at a time, so that from each batch over 20,000 integrated circuits or microprocessors will be produced. As a modest semiconductor company will make more than a hundred batches in a week, its total output will be in the region of 100 million complex circuits per year. This staggering throughput is achieved because microelectronics manufacturing is a mass-production industry where many batches of wafers are produced at once. It may cost anything from $500 to $1000 to produce a wafer but there may be two hundred or more perfectly working integrated circuits on each.

In a production environment the processing time may take up to one month. Considering that there may be many batches of silicon wafers, of different circuit functions, in the line at one time the scheduling and planning operation is very detailed. A complete integrated circuit process usually involves more than a hundred individual steps, all of which must be performed identically on each silicon wafer, during each production shift, on each successive day. The devices are tested after production on a tester that grades them into various performance categories. The percentage of good working circuits relative to the number of potentially working candidates is referred to as the yield, which provides an important indicator of the financial success of the semiconductor plant. Basically, the overall yield depends on the area of the circuit and ranges typically from above 99 per cent for a very simple chip of several thousand transistors and perhaps a few square millimetres in area, to less than one per cent for a complex LSI circuit. The yield is never 100 per cent because each operation in the fabrication sequence will vary slightly and produce a variance in the final device performance. A typical example is the condition of the photomasks used to define the patterns in the wafer fabrication process. The masks become worn and will be degraded by dust particles in the environment. Strict quality control must be operated, and the processing environment must be controlled with respect to dust particles, and the temperature and humidity of the air. Thus to achieve low cost per integrated circuit function, it is necessary to maintain high yield in processing and achieve high wafer throughput. Due to mass production the cost per transistor on the chip has fallen by a factor of over 100 from 1971 to 1979. Microelectronics is thus the world's finest inflation beater. This, of course, is the reason why the microelectronics chip is

of silicon technology. A modern integrated circuit can have more than 100,000 transistors, all of which nestle closely to one another. This means that each device can potentially interact with its nearest neighbours, because of the small spacings involved. Fortunately, each designer is normally supplied with a set of design rules that specify minimum spacings between individual components, so that he is then able to choose the required interconnection configuration for each of the required number of transistors, and then lay out a circuit plan of the complete chip – designers call this 'defining the topology'. Because of the enormous complexity of the problem faced by each designer, he needs to employ computer aids wherever possible and, in addition, it is common for designers to work in teams to reduce the design time. For example, a microprocessor LSI design would typically consume over ten man-years of effort, which implies a team of about ten men because of the speed of innovation necessary today to keep up with world competition.

Thus an important fixed cost in integrated circuit development is the design cost. Because circuits are so complex, it may take a designer or a team of designers between six months and about two years to design a circuit. In addition to this identifiable labour cost is the capital cost associated with design aids. To be effective, the designer must have at his disposal a dedicated design computer to help with the checking of the expected circuit performance and to assist in the layout of the transistor structures. It is very common nowadays for a design team to require computer design aids totalling $1m, and it is expected that as the VLSI-era unfolds, investment levels ten times this amount per team will become commonplace.

The overall design cost can be essentially removed if the systems engineer, who requires a supply of integrated circuits, is able to purchase standard off-the-shelf chips. Here the initial design cost can be amortised over a vast production run of perhaps several million devices. Where the customer cannot buy such pre-designed circuits, a custom design must be commissioned. This will ultimately offer a lower cost per function than the off-the-shelf or catalogue approach and, in addition, will offer performance advantages such as reduction in system size by using one custom package rather than several catalogue circuits. The designer is increasingly required to optimise the minimum cost per system-on-a-chip, rather than the minimum cost per individual electronic function. This change in emphasis over the past few years means that even more than previously the designer has an absolutely crucial role to play.

Microelectronics enables existing electronic systems, for a vast range of applications, to be produced at minimum cost. The enormous power of the microelectronic chip now enables hitherto unexplored applications, for example, in the sphere of mechanical and chemical engineering, where electromechanical control was previously employed. The integrated circuit currently accounts for nearly fifty per cent of all electronic components, with some circuits being truly integrated systems. The new technology is all-pervasive and the major fields of application are computers, calculators, industrial control, electronic instruments, medical electronics, aircraft and space applications, communications and consumer electronics, and this is by no means an

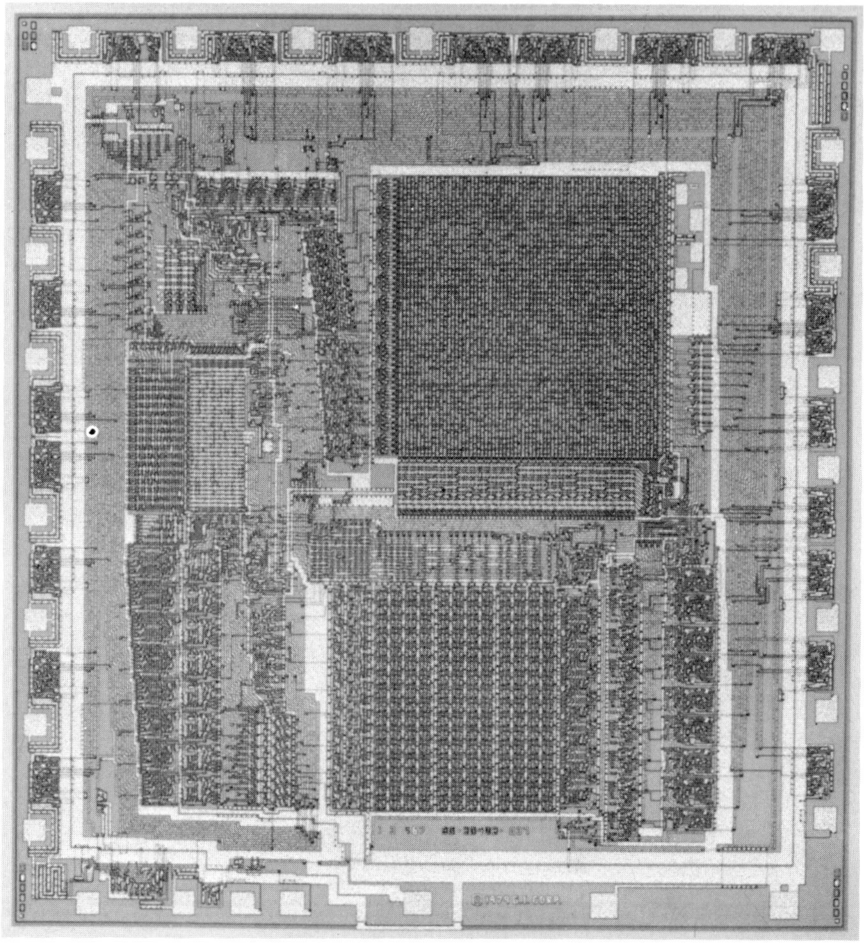

Microprocessor chip. The microprocessor performs the basic functions of a computer. This integrated circuit microprocessor measures approximately 5 mm square and is an 8-bit processor with a 12-bit instruction word, a 512 × 12-bit read-only memory and 32 × 8-bit registers. The integrated circuit microprocessor demonstrates the significance of microelectronics technology by allowing highly complex circuits to be realised on one chip, which can be applied to a wide variety of applications by means of software programming. *Photograph courtesy of General Instruments Microelectronics Ltd.*

exhaustive list. There will, as a result, eventually be an enormous change in lifestyles through the impact of the silicon chip, and a dozen or so books on this topic have already been written, most of which have the word revolutionary in the title. However, what most people fail to appreciate is that microelectronics can perform many of the less exciting jobs which many human beings have to fulfil today and that microelectronics, through robots, can remove much human drudgery. However, the emotive issue is the question of who will control the microelectronic robots and production lines. This is a fair question to ask because the microchip may be regarded as a small electronic brain and, although its capabilities do not compare favourably with those of the human brain, it is anticipated that by the end of the century its computing power will exceed that of the human brain, although the debate over whether actual intelligence is present will probably remain. Undoubtedly, the most significant advance in integrated circuits in the past few years has been the development of the microprocessor, which is a small computer with computational logic and digital data storage fabricated on a single LSI chip. The microprocessor has already led to many extremely significant developments where an electronic brain is required to control the functions of a small system. Significantly, the cost of these microprocessors is already relatively low, with projected costs of tens of dollars in the near future, thus enabling them to be used in many mundane tasks including washing machine control, toys, traffic programmers, and car instrumentation.

Integrated electronics has now penetrated virtually every field of human activity and developments in device design and technology are still progressing so rapidly that it is impossible to appreciate the extent to which these developments will affect our daily lives. It is now true that integrated circuits can satisfy most electronic systems needs in terms of component complexity and performance: the challenge is now to select the appropriate microcircuit technology, and then to engineer the system for minimum system cost.

2. MICROELECTRONIC CIRCUIT FABRICATION

At a time when most sectors of industry are striving hard to achieve growth rates of a few percent, the electronic components industry has repeatedly demonstrated an ability to produce growth figures of 15 to 25 per cent per annum throughout the world. On a more restricted basis, companies at the forefront of microelectronics technology show even more spectacular growth rates. Added to this sales growth is an increase in the functional complexity of integrated circuits of around 100 per cent per annum. This has produced, for little extra manufacturing cost, a dramatic fall in the net production costs of the basic building blocks of electronic systems, and it is this feature that has created the new markets for microelectronics.

If we now look behind the facade of publicity and try to analyse the factors that make the development and application of semiconductor technology so unusual, we find that there is no unique invention but rather a highly efficient blend of many disparate sciences and technologies, which have been combined together to give a uniquely successful production process. In the twenty years of its existence the integrated circuit has been the centre of intense process technology development, and this in turn has permitted more ambitious circuit designs and new applications. The resulting successes and corporate growth have in their turn justified still more process development, thereby producing a high-technology spiral of development that shows little sign of abatement.

Modern electronics is having a serious impact on our everyday lives but, although the range of applications seems to be never-ending, few radically new concepts are involved. The heart of the 'silicon revolution' lies in the fact that the production process for manufacture of integrated circuit chips has proved capable of radical development year after year. The result is an engineering product that is unlike any other and whose evolution over a twenty-year period may be summarised by zero increase in basic chip costs, five orders of magnitude increase in transistor density per chip, and six orders of magnitude reduction in transistor area. This present chapter considers how integrated circuits are manufactured and demonstrates the main features of the technology that has made such dramatic changes possible.

The integrated circuit production process has many variants but its major features are that it is highly capital intensive and intrinsically suited to high-volume production, features that are welcome in an industrial environment. The pace of process development has been so rapid that only the most efficient industries have been able to maintain adequate investment levels to exploit their internal development programmes.

The raw material used as the starting point for almost all integrated circuit production is silicon. Silicon is an element that is capable of being refined to a very high purity and that electrically behaves as a semiconductor. This means that, relative to insulators or metals, its conductivity can be varied either by electrical means or by the addition of small quantities of selected impurities. The latter property is used to produce transistor structures on the silicon surface, which can be electrically switched on or off to give the basic binary components that form the basis of all digital computers.

Silicon boule and wafer. A boule, which is a single crystal of pure silicon, is drawn from molten silicon to form a cylinder some 130 mm in diameter and 1 to 3 m in length from which the individual wafers (0.5 mm thick) are cut. The boule has a flat surface ground along its length to indicate the crystallographic axes so that all the wafers can be oriented in identical fashion during processing.

The production process starts with molten silicon (above 1400°C), which is cooled under controlled conditions to give a perfect single crystal (boule) that may typically be 70–130 mm in diameter and 1–3 m long. Each boule is sawn into a large number of thin wafers or slices, each about 0.5 mm thick, and the wafer surfaces are subsequently cleaned and polished. On each wafer a large number of chips are defined, for example 260 chips each 4 mm square on a 75 mm diameter wafer, and each chip may contain up to 10,000 transistors (more if it is produced by one of the most advanced processes). Since wafers are usually processed in batches of 25–150, the whole production process is intrinsically capable of giving very large numbers of devices at low unit cost.

To the non-specialist, the most significant feature of the integrated circuit fabrication process is its complexity. It may take three to four weeks to produce a batch of circuits and over 400 different operational steps may be required. Fortunately, however, there is a basic process cycle, which is repeated several

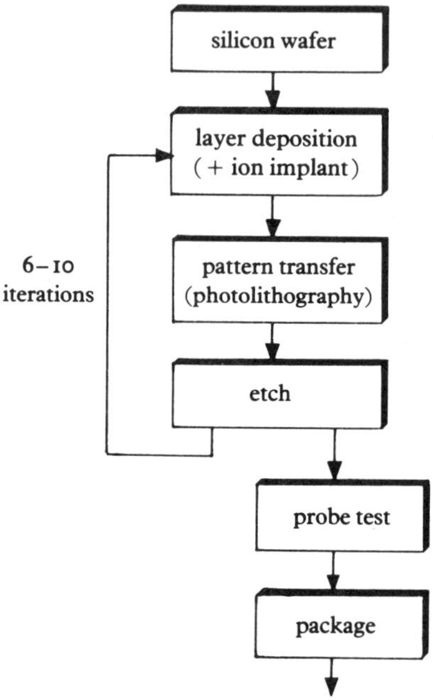

Integrated circuit fabrication sequence. The basic integrated circuit manufacturing process consists of a series of repetitive operations on the silicon wafer. These operations consist of 'deposition' to form the required materials over the complete silicon surface, 'pattern transfer' to define those specific areas of material required in the circuit and finally 'etching' to remove unwanted material. When the process is complete the individual circuits are tested and packaged for use.

times within the overall process sequence. Little has changed in the basic process cycle since the first integrated circuits were produced just over twenty years ago. Transistor structures are built up through many iterations of material deposition, photolithography and etching and, although the structures have become more complex over the years and it may be necessary to pass a wafer through eight to ten photolithography stages, the basic operation remains. A few radically new process steps, such as ion implantation (whereby ions are bombarded into the silicon) and chemical vapour deposition, have been introduced but most emphasis has been given to two specific aspects of process development: reduction of transistor feature sizes and methods of improved process uniformity for higher production yields. It is the successful exploitation of these results that has transformed the simple interconnection of a few transistors in the early sixties into the complex microprocessors of today. The skills of many scientific disciplines have been involved, including optics, chemistry, solid state physics, precision mechanics and metallurgy as well as electronics, and it is the interdisciplinary nature of this technology that has supplied some of the most important stimuli for more advanced development.

Before describing the detailed fabrication of integrated circuit structures, it is constructive to examine some of the individual process steps in detail so that

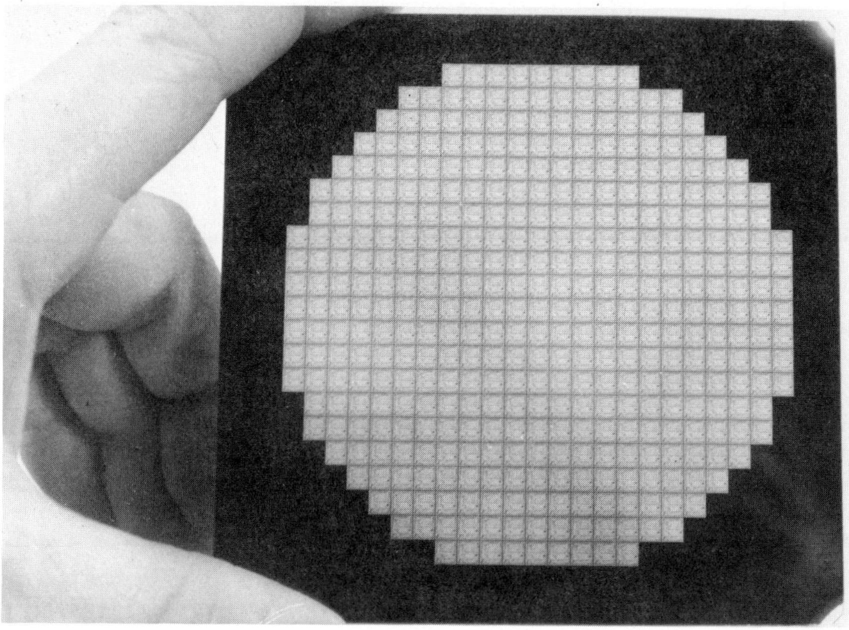

Photographic mask. A photographic mask is used to define areas of the silicon surface that are required in the circuit. One mask is required for each cycle in the process. The mask consists of an array of (usually) identical circuit structures on a glass plate some 70–130 mm in diameter, and it is possible that each mask may be used only once or twice before its surface becomes damaged.

their interaction will be clear. A typical complex integrated circuit may be defined by eight separate layer patterns, each perfectly aligned with the others. In a typical production process, the minimum alignment and registration tolerances must be kept within one micron. To achieve this accuracy a photographic process is used that is, in principle, similar to conventional photography, but the standards demanded of the lenses and emulsions are about one order of magnitude better than those used in normal photography.

The most straightforward way to produce masks is to write full-size patterns either with very fine light sources or with electron beams. This approach offers considerable flexibility and high accuracy but it is expensive, and a more conventional approach uses a pattern generator machine to produce images that are ten times larger than the final size. The pattern generator usually consists of a controlled rectangular aperture through which light is projected onto a photosensitive plate. By varying the sequence of aperture shapes, a mask pattern may be generated. The resulting plate or reticle is typically 5 cm square if the final chip is to be 5 mm square. The second stage of this mask-making process involves photoreduction of the reticle image to a final-size working plate. In this process, the image is then stepped laterally by one chip length and another pattern produced, and so on. This is called a 'step-and-repeat' reduction, and it is used to fill the mask plate (typically 10 cm square) with a regular array of identical chip images.

The one feature that makes silicon unique among semiconductors is that it is relatively easy to grow a layer of silicon dioxide on its surface. This oxide is a good insulator, can be rapidly temperature-cycled to 1000°C or more and, since it uses the native silicon of the wafer, provides the silicon with a stable and passive surface. Oxide layers are usually grown by placing silicon wafers in an atmosphere of oxygen or steam at 900°C or above. The thickness and characteristics of the oxide depend on the exact growth conditions and the deposition time may vary from twenty minutes to fifteen hours. In the latter case a very thick oxide film about 1 μm thick is grown. If the silicon wafer surface is not directly exposed, silicon dioxide can still be deposited by the oxidation of silane gas. Such vapour-phase reactions are very convenient because they do not depend on the chemicals being present on the surface of the silicon wafer. The drawback of the process is that the resultant film thickness tends to vary across a wafer, and this in turn gives poor reproducibility and yield. This problem has been overcome in a very elegant fashion by initiating the required reaction at low pressure, and this approach has an added bonus for production because it allows wafers to be stacked vertically in the reaction tube and high production throughput is possible. Low-pressure chemical vapour deposition (LPCVD) is used to produce layers of polycrystalline silicon and silicon nitride, with the polycrystalline silicon being used as a conductor that is compatible with the substrate material and can withstand high temperature processing, and the silicon nitride being used as an insulator that is not attacked by the etches used to define patterns in silicon dioxide (and vice versa).

Photolithography is the heart of the whole fabrication process, because it provides the means of transferring the required image from a photographic mask plate to the wafer surface. To achieve this the silicon wafer is coated with a thin layer of photoresist, which is a light-sensitive emulsion, applied as liquid. A small quantity is placed on a wafer held on a vacuum chuck. The wafer is spun at about 6000 rpm and the resist spreads to give a very uniform layer about 1 μm thick. The resist is baked gently to dry it and it is then ready for exposure.

There are two types of photoresist and both are widely used. Positive resist dissociates under ultraviolet light, so any exposed areas can be developed and removed. Dark areas on a mask are therefore reproduced as corresponding resist areas on the wafer. With negative resist, ultraviolet light polymerises the resist so that exposed areas are left with a resist coating after development. Positive resist is more tolerant of adverse operating conditions and gives better resolution, so its use is increasing throughout the industry.

As implied by the description of resist operation, the pattern transfer from mask to wafer is achieved by exposure to a high-intensity ultraviolet light source. This may be achieved by a contact print, where the mask and wafer are brought softly into contact with each other, or by an alternative method known as projection printing, which results in less mechanical damage to the

Mask alignment. Since a series of 8–10 masks may be required in a fabrication process, care must be taken to align each mask relative to previous masks because mis-registration between levels would tend to create imperfect circuits. The use of a television camera and monitor attached to the mask aligner facilitates registration to an accuracy better than one micron.

mask and wafer surfaces but which requires more expensive equipment. In either case serious problems arise when the second and subsequent masks have to be aligned to the patterns already on the wafer. The mechanical and optical demands on the equipment to attain alignment accuracies of 1 μm or better are high, and since any excessive mis-registration on any of the mask layers would affect all chips, the penalty for inaccuracy is a scrapped wafer.

When a pattern has been successfully aligned and developed, the remaining photoresist layer is used as a mask to protect some parts of the wafer surface while the surface layer of silicon, silicon dioxide, silicon nitride or metal is etched away. The photoresist layer may then be removed to complete the pattern transfer process. Because photoresists are organic compounds they cannot withstand the high process temperatures used for layer deposition, so

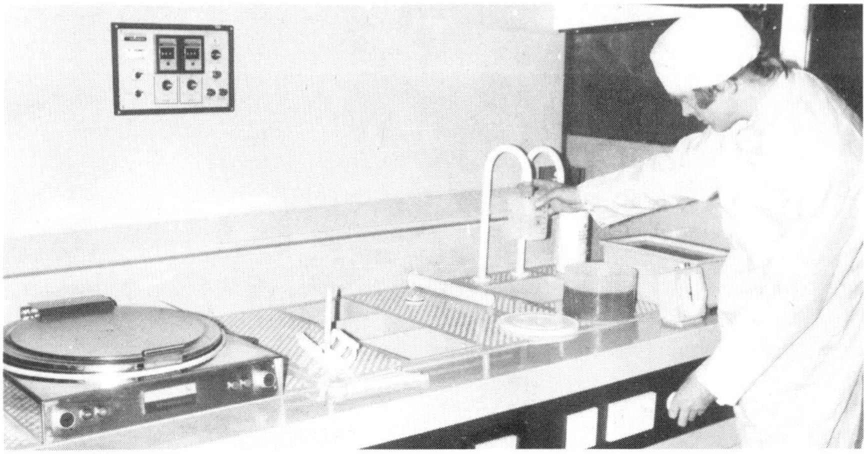

Wafer etch. Subsequent to each of the patterning stages in the fabrication sequence the unwanted material is removed by chemical etching. Here strong acids are used to remove the material, and the wafers are then cleaned in high-purity de-ionised water and dried prior to the next deposition operation.

they cannot be used as part of the active device structure.

During the lithography process wafers are very vulnerable to airborne dust contamination. A speck of dust is large by comparison with the typical device features being defined, so any dust that lands on a resist layer could give rise to a defect that would produce a non-functional chip. It is therefore normal practice to carry out all lithography operations in filtered air. Temperature and humidity control and yellow safe-lights are also usually provided.

A silicon surface covered by a protective layer of oxide tends to accumulate a small quantity of negative charge, which is termed an *n*-type surface layer. In a common type of transistor used in integrated circuits, the device operates by controlling the flow of negative charges (electrons) between surface electrodes. This would be a reasonable arrangement for an isolated transistor on a chip, but in the presence of adjacent devices uncontrolled currents would also flow through the *n*-type surface layer, rendering any circuit interconnection between devices meaningless. To overcome this problem, ions with a net positive charge are implanted through the oxide layer to neutralise the natural *n*-type surface and provide a slight excess positive charge. The resulting *p*-type surface provides isolation between adjacent devices.

The implantation machine is an ion accelerator. It consists of a long evacuated tube with a source chamber at one end. Ions are extracted from the source at up to 40 kV and then pass through an analysing magnet. The ion beam is deflected by the magnetic field and only those ions with the required energy (and mass) pass through the magnet chamber. The final acceleration stage provides high-energy ions (up to 200 kV) that impinge on the silicon wafers, which can be either rotated in front of a fixed ion beam or scanned by an electrostatically-deflected ion beam. The depth of ion penetration depends on

Ion implanter. An ion-implantation machine provides the means of injecting very accurate doses of impurity atoms or ions into semiconductor surfaces. The ion energy determines the depth of penetration and the ion beam current sets the total charge implanted. The technique gives the accurate and controlled changes in surface conductivity that are needed for fabrication of complex small-geometry transistors.

their energy but it is typically 0.5 μm. The ions may therefore penetrate a thin layer of oxide but any part of the chip can be masked by a one-micron layer of photoresist. Since ion implantation is a low temperature process where the ion distribution and total dose can be separately controlled it has very significant advantages for any production process. The only drawback is that implantation machines are expensive ($500,000) and complex to run and maintain. This serves to emphasise one of the basic problems of all advanced integrated circuit fabrication – in addition to the formidable technical specifications to be met, the organisational problems involved in the management of a process line are considerable.

Ion implantation can also be used to establish the characteristics required for selected individual devices. Basically, two types of transistor are used in integrated circuits. In the first type there is normally no current flow between its two main electrodes (called the source and drain) until the control voltage applied to the gate electrode turns the device on; this is called an enhancement-mode device. The other type of transistor is normally on (conducting) and can be turned off by a control voltage. It is called a depletion-mode device and is principally used as a simulated high-value load resistor. Implantation of the

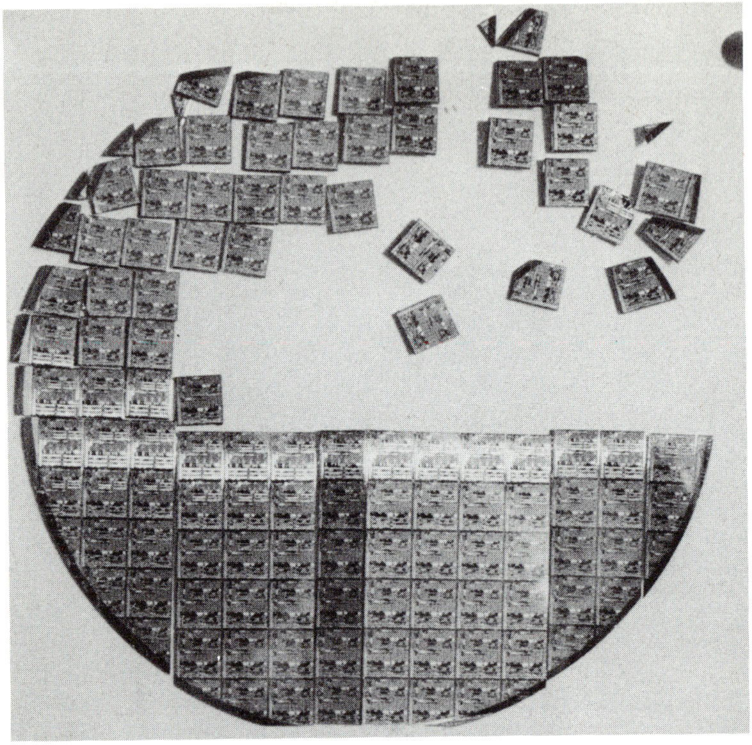

Processed wafer. The finished wafer contains an array of identical circuits that can be probe-tested prior to cutting with a fine diamond saw to yield the individual circuits.

appropriate ion species between source and drain defines the transistor types and the required threshold control voltages. Implantation is clearly a crucial process step since it determines the operating device parameters required in the circuit design, thus uniformity and reproducibility are the prime performance requirements for any ion implantation machine.

When all the transistor structures have been defined on the silicon surface, the final process step is to interconnect them in the required circuit configuration. To do this, the standard cycle of deposition-lithography-etch is used. In most cases the metal is aluminium and it is deposited by evaporation. Under high vacuum aluminium first melts, then evaporates at about 800°C. The vapour condenses throughout the chamber and on the wafer to give a layer about 1 μm thick. By the time a wafer is ready for metallisation the plane surface of the starting wafer has been converted into a surface with many bumps and depressions – at least on a micron scale – and to ensure that wafers are evenly coated with metal with no cracks or breaks at surface steps, the wafers are rotated during the evaporation process.

The metal layer also defines the relatively larger bond pads around the perimeter of each chip. When the fabrication process is complete each chip is tested with an automatic probe test system, non-functional devices are marked, and the whole wafer is sawn up into individual chips. The non-functional devices are discarded and each remaining chip is mounted in a package and a gold or aluminium wire, typically 25 μm in diameter, is welded between a bond pad on the chip and the package leadframe. This bonding operation is a very labour-intensive process.

For twenty years the packing density achievable in integrated circuits has almost doubled each year, as defined by Moore's Law. For most digital computing functions it is only necessary to recognise when a transistor is switched on or off, and the physical size of the device or the current flowing in it are not relevant to the function required. It is therefore possible to reduce the feature sizes of individual transistors in a circuit and so achieve a higher component packing density on each chip. Based on the possibility of reduced feature sizes, how far might we expect to be able to extrapolate Moore's line? Advanced production processes today offer three-micron features. By fully exploiting known production techniques, the potential exists to reduce features to values approaching 0.5 μm by the end of the decade. Beyond that predictions are uncertain, since with 0.5-μm devices oxide layers are so thin that they no longer behave as insulators and the carrier transport process in silicon changes. Further, power dissipation becomes a dominant problem.

To provide a measure of the potential offered by such reduced feature sizes the complex eight-bit microprocessors that are currently being widely used have about 6000 gates (or basic logic circuit blocks) on each chip. By 1990 we might expect to have 600,000 gates per chip and the prospect of an increase of about 100 times in computing or memory power for little extra hardware cost. To date, the pace of development in microfabrication has been maintained by the knowledge that no matter how complex the integrated circuit might be, it could be used. In the future we can be less sure. The level of design effort and systems organisation required to exploit the available circuit complexity will seriously limit the pace of development unless new approaches are found to solve these problems. However, although there are discernible physical limits to microelectronics technology, it seems likely that for the foreseeable future the only serious limitations will be set by our own imaginations.

MOS integrated circuit fabrication sequence. The main processing steps involved in the MOS integrated circuit fabrication sequence involve deposition (implantation) to form new materials on the silicon surface, photolithography to define specific areas of interest, and etching to remove all unwanted materials. For clarity, only five stages of photolithography (mask levels) are depicted here for the basic *n*-channel MOS process. Actual processes involve an additional three or four mask levels.

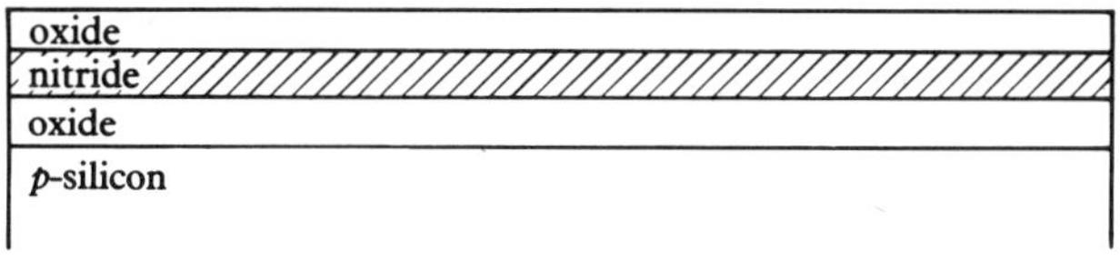

1. **Initial clean.** Removal of any organic material and surface oxide in acid and hydrogen peroxide.
2. **Initial oxide.** Production of thermal oxide layer, 0.05 microns thick.
3. **Silicon nitride.** Deposition of a thin film (0.05 microns) by a process called low-pressure chemical vapour deposition (LPCVD).
4. **Masking oxide.** Growth of an oxide layer by thermal oxidation of silicon on the silicon nitride surface.

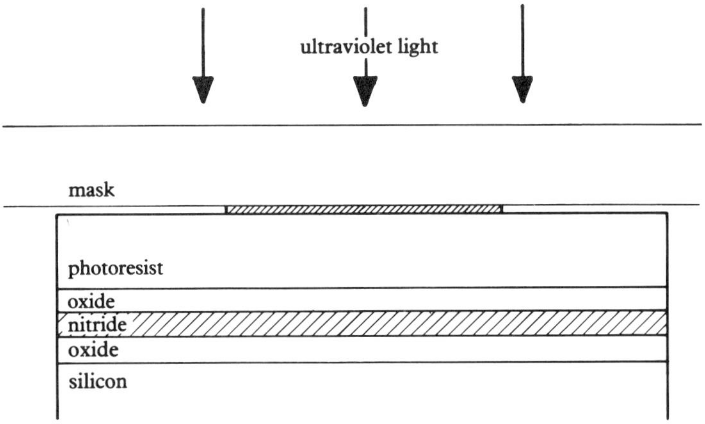

5. **Photolithography.** Definition of all the regions required for active devices such as transistors.

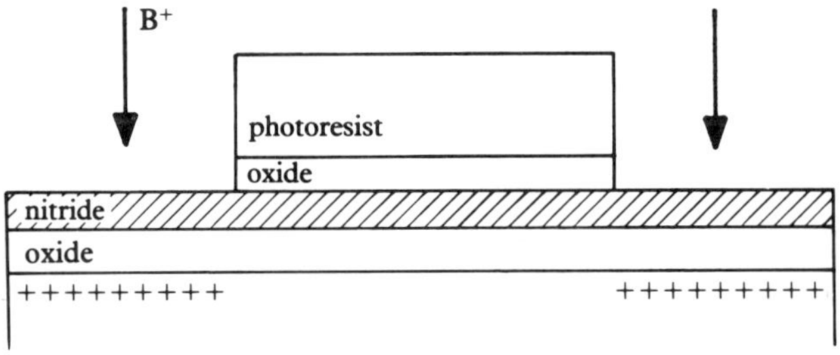

6. **Boron implant.** Implantation of boron ions into the silicon surface. The energy of the boron ions is such that they can penetrate the oxide–nitride–oxide sandwich but cannot penetrate the active regions, which are masked by the photoresist layer. Boron creates a net positive charge at the silicon surface between the areas that will become transistors, thereby providing electrical isolation between devices.
7. **Oxide etch.** Removal of the thin surface oxide layer except where the photoresist coating remains.

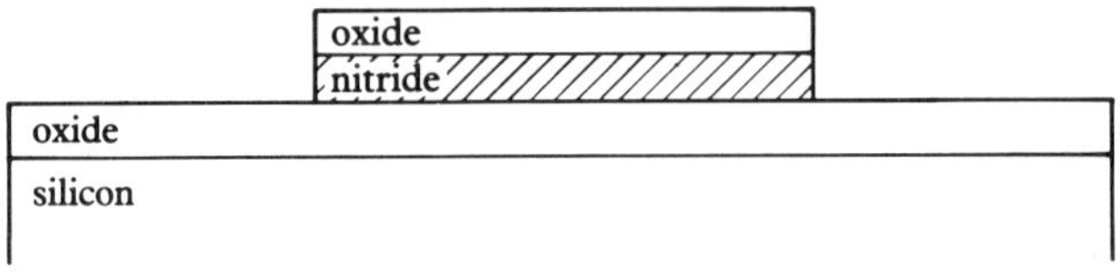

8. **Resist strip.** Removal of photoresist in a strong oxidising agent such as oxygen plasma or fuming nitric acid.
9. **Silicon nitride etch.** Etching of silicon nitride in all areas not protected by the surface oxide. Photoresist cannot be used because it is attacked by the nitride etch chemicals.

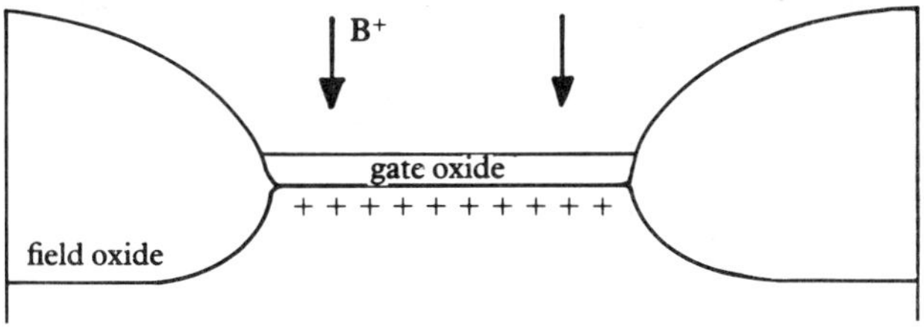

10. **Field oxide.** Oxidation of silicon on the wafer surface in steam yields a layer of oxide one micron thick in all areas not protected by nitride. In nitride areas the resulting oxide thickness is only about 0.01 microns.

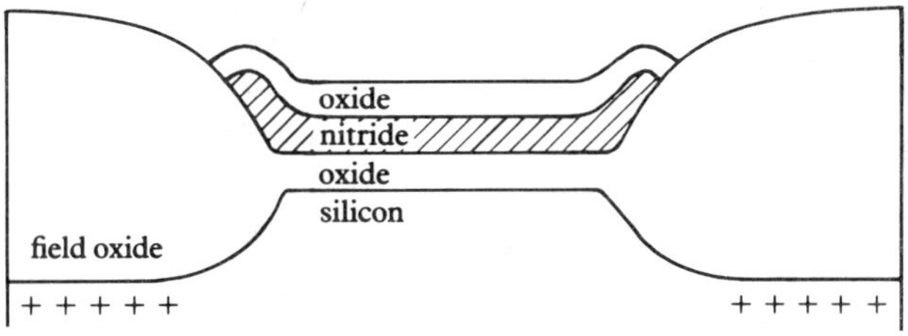

11. **Oxide etch.** Removal of the thin surface layer of oxide produced on the nitride surface. This operation has little effect on the thick field oxide layer.
12. **Nitride etch.** Removal of the nitride layer.
13. **Oxide etch.** Removal of the thin oxide remaining in the regions that define the positions of active components.
14. **Gate oxide.** Growth of a very pure oxide film approximately 0.08 microns thick. This layer forms the basis of the MOS devices and its uniformity, purity and stability are critically important.
15. **Boron implant.** Implantation of boron ions to define the operating voltage (threshold voltage) for enhancement-mode MOS transistors.

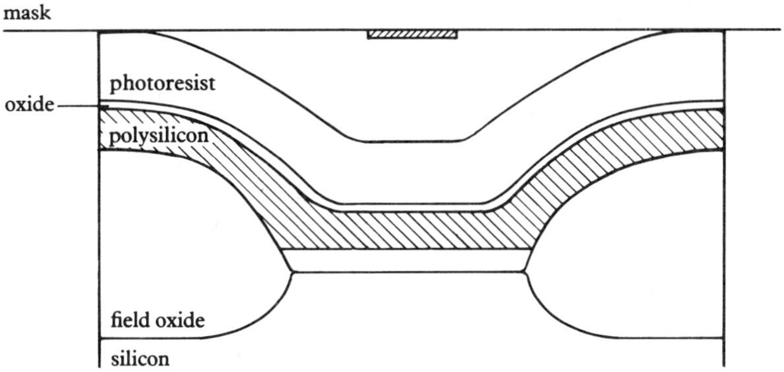

16. **Resist strip.** Removal of any surface contamination produced by organic materials present in the ion implanter system.
17. **Anneal.** Ion bombardment damages the silicon crystal structure. Annealing permits local atomic repair of the surface and removal of implantation damage.
18. **Polysilicon deposition.** Use of an LPCVD process to create a polysilicon layer 0.35 microns thick. The polysilicon will be used as conductors and gate electrodes for transistors.
19. **Oxidation of polysilicon.** Production of a thin surface layer of oxide by thermal oxidation.
20. **Photolithography.** Definition of the polysilicon areas that are required for gate electrodes or interconnections between devices.

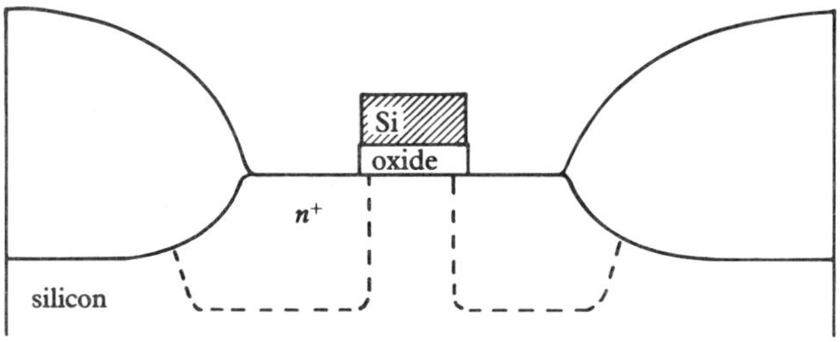

21. **Oxide etch.** Removal of surface oxide in all regions not protected by photoresist.
22. **Resist strip.** Removal of photoresist from polysilicon areas.
23. **Polysilicon etch.** Removal of polysilicon in areas where it is exposed – areas not protected by the surface oxide layer. As with step 9, photoresist cannot be used as an etch barrier because it is attacked by the polysilicon etch.
24. **Oxide etch.** Removal of oxide from the top surface of the polysilicon and from regions between the polysilicon and the field oxide.
25. **Phosphorus deposition.** Diffusion of phosphorus into the polysilicon layer to increase its conductivity. The phosphorus also diffuses into exposed regions of the silicon substrate, thereby defining the other two electrodes of the MOS transistor – the source and drain regions.
26. **Phosphorus de-glaze.** Removal of the phosphosilicate glass grown during step 25. This material has a low melting point and is a relatively poor insulator.

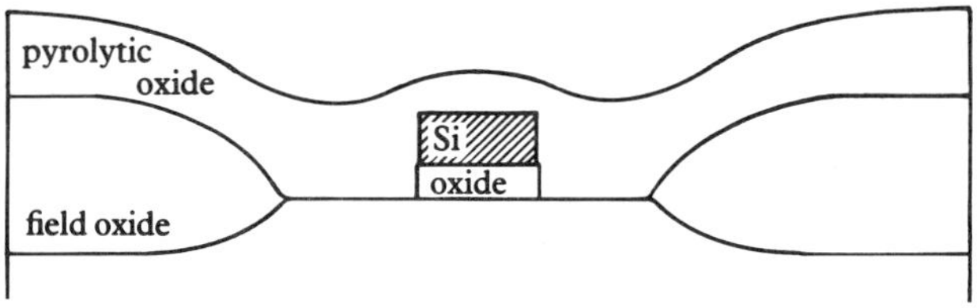

27. **Polysilicon oxide.** Growth of a layer of dense oxide 0.2 microns thick to provide a good base for step 28.
28. **Pyrolytic oxide deposition.** Production of a uniform, well-defined phosphosilicate glass layer about 0.75 microns thick by means of phosphine gas.
29. **Reflow.** Softening of the glass layer to smooth the sharp poly edges for improved reliability.

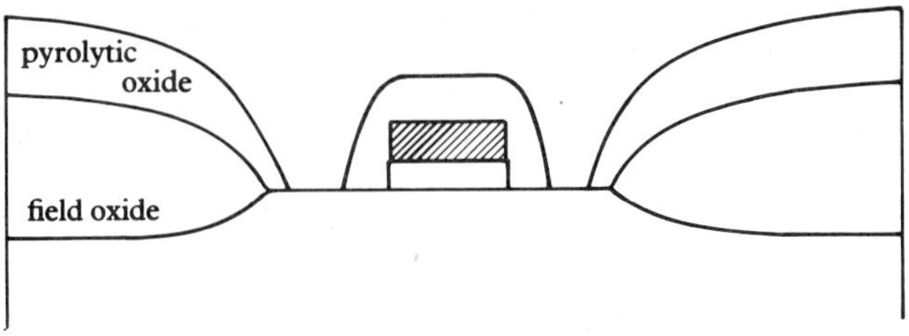

30. **Photolithography.** Definition of areas where contact will be made to polysilicon or diffused regions.
31. **Reflow etch.** Removal of reflow oxide from areas not protected by photoresist to expose contact windows.
32. **Resist strip.**
33. **Second reflow.** Rounding edges of defined contact holes.

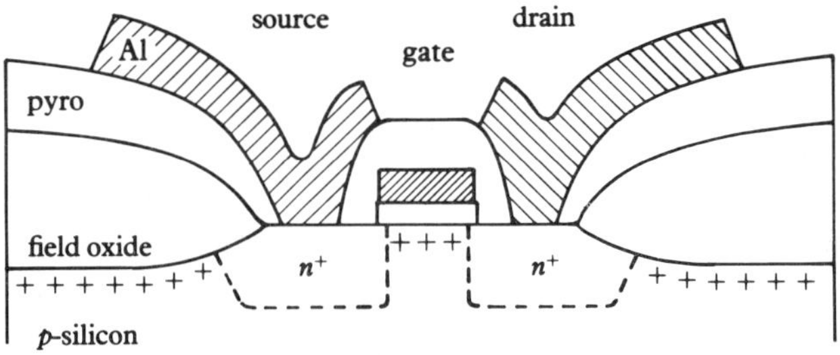

34. **Aluminium deposition.** Evaporation of a layer of pure aluminium 1.5 microns thick over the whole wafer surface.
35. **Photolithography.** Definition of aluminium interconnect pattern.
36. **Aluminium etch.** Removal of aluminium from all areas not protected by photoresist.
37. **Resist strip.**
38. **Sinter.** Annealing at low temperatures to provide good contact between silicon and aluminium layers.
39. **Pyrolytic oxide.** Deposition of a layer of oxide 0.75 microns thick over the whole wafer to provide mechanical/chemical protection of the finished circuit.
40. **Photolithography.** Definition of oxide windows over bonding pads.
41. **Pyro etch.** Removal of pyrolytic oxide over bonding pads to expose metal pads.
42. **Resist strip.**

3. MICROELECTRONICS AND THE MICROPROCESSOR

Public interest in the microchip has been stimulated in recent years by the extensive exposure that this new technology has received in the media. The chip has been presented, on the one hand, as a solution to all our economic ills and, on the other, as a potential cause of severe unemployment. The remarkably rapid technical progress of microelectronics technology has tempted many writers to sensationalise the subject, giving the impression that the chip was a sudden and revolutionary invention. In fact, the chip (or integrated circuit) has evolved in a continuous way over the past three decades and the microprocessor as we know it today is merely one example of microelectronics technology, a milestone in an evolutionary process that began in the fifties and which is widely expected to continue into the foreseeable future. We need to place the microprocessor in its proper perspective, both historically and in relation to other developments in microelectronics and to its applications in electronic and computing equipment.

The first microelectronics devices, discrete transistors and diodes, began to replace thermionic valves shortly after the end of the Second World War. It was soon realised that the methods used to fabricate these discrete devices could equally well produce complete integrated electronic circuits comprising several components embodied in a single semiconductor chip. An integrated circuit could be fabricated much more cheaply than the corresponding conventional circuit, in which several discrete components had to be connected together. Integrated circuits also used less power and proved to be more reliable than conventional circuits and hence they enabled increasingly sophisticated electronic equipment to be produced at low cost. Equipment manufacturers quickly recognised the cost advantages of using integrated circuits and there was a growing demand for more complex functions to be made available as single-chip devices. The semiconductor manufacturers were able to respond to this demand because the large market for their products justified the necessary heavy investment in research and development and in sophisticated production equipment. The first small-scale integrated circuits (SSI) were introduced in the early sixties. They comprised small sub-systems, such as logic gates and flip-flops, on a single chip incorporating between ten and a hundred transistors. More complex devices, such as 128-bit shift registers and complete four-bit adder circuits, became available by the end of the decade. These devices embodied hundreds of transistors and were described as medium-scale integrated circuits (MSI). The price of integrated circuits fell dramatically over this period as a consequence of the capital-intensive nature of the semiconductor fabrication process, which allows devices to be produced at extremely low marginal costs in high-volume production. Indeed, the cost of the integrated circuits themselves came to represent only a small proportion of

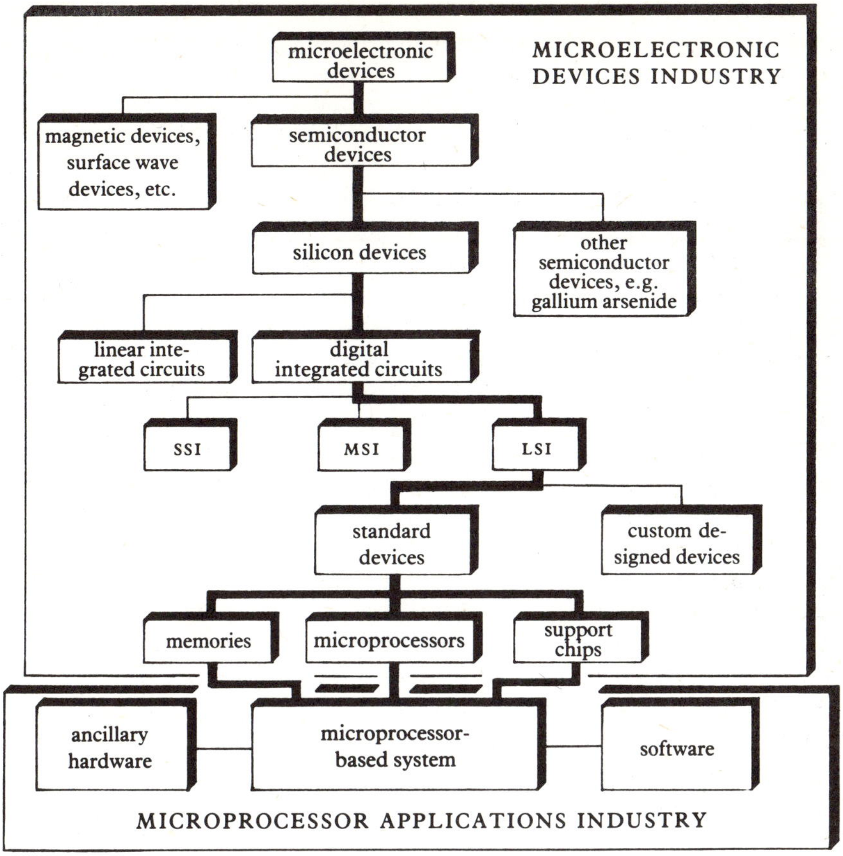

Microelectronic devices. Microprocessors are only one of a wide range of devices produced by the microelectronics industry. The other devices offer alternative, and sometimes better, ways of implementing systems and custom-designed chips in particular could become more attractive than microprocessors in the future. A clear distinction exists between the microelectronic devices industry, which is concerned with the mass production of microprocessors and other devices, and the microprocessor applications industry, which employs microprocessors and other devices, together with software, in order to implement microprocessor-based systems.

the total cost of a typical piece of electronic equipment. Other costs, such as assembly, printed circuit boards, interconnection, power supplies, mechanical design and testing became dominant. These assembly and related costs depend largely on the number of separate integrated circuit devices employed in the equipment, the so-called package count, so that the cost of the equipment is greatly reduced if several separate devices can be replaced by a single, more complex chip.

By the early seventies, however, it was realised that the increasing complexity of integrated circuits would bring problems as well as benefits. Advances in microelectronics technology had been so rapid that it had become possible to fabricate large-scale integrated circuits (LSI) that incorporated around 2000 transistors on a single chip, an order of magnitude more complex than the typical MSI device. The development of a new LSI circuit requires many man-years of highly skilled man-power supported by expensive computer-aided design equipment, and these design costs can only be justified by high-volume sales of the chip. It is unfortunate, however, that, in contrast with MSI devices (which may be regarded as components or sub-systems, appropriate to a wide range of applications), the functional complexity of an LSI circuit tends to restrict it to a specific, narrow application area. Purpose-designed LSI chips are therefore only economic for very high-volume applications such as telecommunications and consumer products such as TV games and digital watches. In order to apply LSI technology to small- and medium-volume applications it was necessary to make available a general-purpose device that could be used in many different situations. This was the microprocessor – a standard circuit that could be customised for specific applications by means of the software programming methods that had been widely used in the computing field for many years.

The first microprocessors were thus conceived as logic-replacement devices rather than as computers in the strict sense of the word. Nevertheless, these early devices bore many similarities to conventional computers, being essentially a simple computer central processing unit (CPU) implemented as a single integrated circuit. Like a conventional CPU, the microprocessor comprised a number of registers and an arithmetic/logic unit together with instruction decode logic and control circuitry. In operation, the CPU fetches instructions sequentially from an external memory (an instruction is a binary word, typically eight or sixteen bits long) and this instruction is decoded by logic circuits in the CPU. The CPU control circuitry then causes the arithmetic/logic unit to carry out operations, such as addition, on the data contained in the registers, thereby implementing the function required by the specific instruction that is being executed. The same, very general principles of operation apply to conventional mainframe and minicomputers, although the early microprocessors were very primitive in their computing power.

The operations that the CPU performs on the data are determined by the instructions, which are stored in a suitable memory device. A complete set of instructions comprises a program. An important difference now emerges between the logic-replacement microprocessor and the conventional computer. The latter is usually required to execute many different programs, for different users. Each user must be able to write his own program to carry out a particular computation, and to enter the program into the computer, which then executes it and prints out the results. The program will then be replaced in the computer memory by an entirely different program, belonging to a different user. The computer CPU is connected to a read/write memory by a data highway or bus. Input/output (I/O) devices are also connected to the data bus and fulfil the essential function of allowing the programs and data to be entered into the machine and results to be obtained from it, via peripheral devices such as the familiar visual display unit (VDU). The read/write memory is used both for storing the program and also as an additional store for data used by the program, since there is usually a requirement to handle more data than can be held in the limited number of CPU registers.

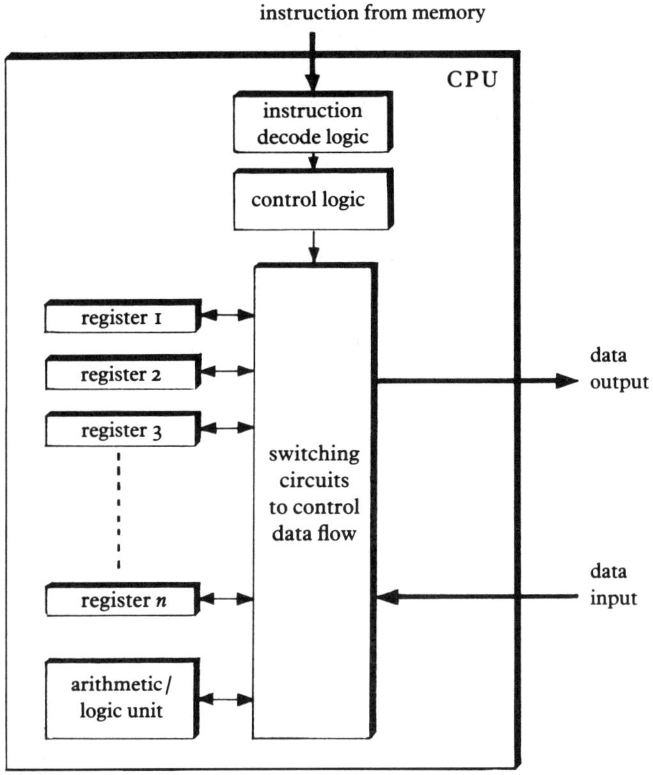

The computer central processing unit (CPU). The arithmetic/logic unit operates on data contained in one or more of the CPU registers and the result is stored in another register. For example, the number stored in register 1 might be added to the number stored in register 3 and the sum returned to register 4. The precise operation (e.g. add, subtract, logical shift, etc.) and the source and destination registers are specified by the instruction that the CPU fetches from memory. The instruction is decoded by the CPU and control circuitry establishes the appropriate data routes to implement the required function.

Logic-replacement microprocessors are used in a rather different manner from conventional computers. The microprocessor is merely a method of implementing a desired logic function, alternative to the use of MSI or custom-designed LSI circuitry. The program is determined by the requirements of the application and forms an integral part of the design of the equipment as a means of customising a general purpose LSI device for the specific application. The user would not be expected to modify the program, any more than he would be expected to modify the internal connections between components;

these features are a matter for the designer of the equipment. It is usual therefore for microprocessor systems to employ two distinct types of memory, read/write and read-only. The program is stored in read-only memory (ROM) which may be read by the CPU in order to fetch the instructions for execution. The program is written into the ROM when the equipment is manufactured and cannot subsequently be altered. Microprocessor systems usually employ some read/write memory for the temporary storage of data but the read/write section of memory is typically much smaller than the ROM program store. The advantages of using ROM for program storage are that ROM is non-volatile (i.e. its contents are not destroyed when the equipment is switched off) and it is also cheaper than read/write memory. Read/write memory is usually referred to in the microcomputer business as RAM (random access memory). The terminology is somewhat confusing, however, because read-only memory (ROM) is also accessed randomly.

Following the introduction of the first logic-replacement microprocessors, it was very quickly realised that the rapid advances being made in microelectronics technology would soon allow the implementation of a much more sophisticated CPU on a single chip, and thus the possibility arose of producing a genuine general-purpose computer at very low cost. A trend towards smaller, less powerful, but cheaper computers had been established for many years in the computing industry. Minicomputers had found wide acceptance in applications such as industrial process control, and these machines represented an increasing proportion of the industry's business, alongside the traditional large mainframe machines, which are best suited to complex data-processing applications. Microelectronics technology would allow the continuation of this trend so that very small businesses, or even private individuals, would be able to afford their own personal computer. The personal computer became a reality by about 1974 with the availability of the so-called second-generation microprocessors. These devices had more sophisticated internal circuitry and their logical structure (or architecture) approached more closely that of conventional computers, although they were still much less powerful than existing minicomputers. The second-generation microprocessors employed eight-bit wide data buses, so that data could be handled as eight-bit words, in contrast to the earlier four-bit architectures. The data word length is often used as an indication of the computing power of a processor because a longer word length allows more flexibility in the instruction set of the microprocessor, especially in respect of the variety of addressing modes that may be employed for accessing data from memory. It also allows greater precision in the representation of numerical values, since an eight-bit word permits representation of 256 values whereas a four-bit word can only represent 16 values. An eight-bit word is especially useful when alphanumeric data has to be handled, because eight-bit binary codes are commonly used for this purpose. It is true that any computation can be carried out by even the most primitive CPU at the expense of more complex programming; the advantage of the more sophisticated architecture is not only that it simplifies programming but also that it improves the speed of computation, because each instruction is more powerful.

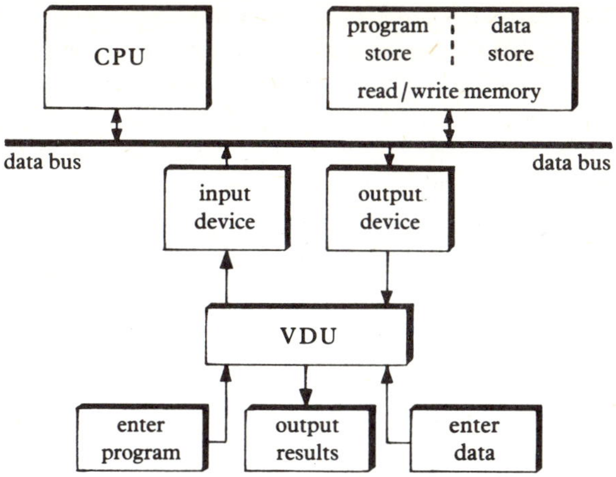

Basic microprocessor structure. In a general-purpose computer the program is written by the user and entered into the machine where it is held in the read/write memory, together with initial data that may also be input by the operator. The program is then executed, i.e. the data are processed and the results are either stored in memory or output. The program may then be overwritten in memory by quite a different program, perhaps belonging to another user. Microprocessors are sometimes used for data processing in this way, but applications in which the program is dedicated are also common.

The typical eight-bit second-generation microprocessor was sufficiently sophisticated to function as the CPU of a general-purpose computer, and many examples of low-cost microcomputers, personal computers or single-board computers soon became available. The lowest priced systems of all are the microprocessor evaluation kits, which usually comprise the microprocessor CPU itself, a small amount of memory, and fairly primitive keyboard and display devices for entering programs and data and displaying results. At the upper end of the range are the more sophisticated personal computers, which are an order of magnitude more expensive and incorporate large read/write memories, full alphanumeric keyboard, visual display unit, floppy disc memory and extensive software support. The advent of these systems stimulated the hobby computing boom as well as making computing a practical proposition for small businesses and schools.

The major limitation of these microprocessor-based systems is their slow speed of computation, and their success was a consequence of penetrating new market sectors in which low cost was much more important than high operating speed. The slow speed of the microprocessor greatly restricted its usefulness in the conventional minicomputer market, however, and an alternative way of exploiting the cost advantages of LSI technology was sought. The outcome was the so-called bit-slice microprocessor, employing bipolar integrated circuits that are inherently very fast. The disadvantage of bipolar technology is that it is not possible to implement high-complexity circuitry on a single chip. The philosophy of the bit-slice microprocessor is that a cross-section of a CPU is implemented as an integrated circuit, a complete processor being assembled by interconnecting several identical integrated circuits. The user of bit-slice devices has, in effect, to design his own CPU architecture, and these devices are therefore of greatest interest to the specialist computer manufacturers. Almost all new minicomputer designs in recent years have been based on bit-slice microprocessors.

It was clear by the mid-seventies that two distinct areas of application existed for microprocessors. These were as logic-replacement devices and as low-cost general-purpose computers. The evolution of microprocessors has now divided into two parallel routes, corresponding to the rather different requirements of these different types of application. The main requirement for the general-purpose computer market was for further improvements to CPU architectures and for longer data word lengths (16 or 32 bits). General-purpose machines are often used to execute lengthy and complex programs, which can only be efficiently written in high-level programming languages such as PASCAL or ALGOL. Programs written in a high-level language have to be compiled (or translated) into instructions that can be directly executed by the machine. The primitive architectures of the second-generation microprocessors made compilation difficult, and compiled programs were often inefficient to execute. The architectural features that facilitate high-level programming were well known from early experiences with minicomputers, and the semiconductor manufacturers were able to incorporate these improvements into the third-generation microprocessors, which were introduced in the late seventies. Further improvements in computing speed were obtained, largely as a consequence of the more efficient handling of data by the more sophisticated architecture.

The rapidly increasing sophistication of the first three generations of microprocessor chips was impressive from the technological point of view but, surprisingly perhaps, there was very little originality in the architectural development of these devices, all of which followed closely upon established minicomputer practices. A very recent device, the Intel IAPX432, announced in 1981, may well be regarded as the forerunner of a fourth generation of microprocessors in which the full potential of microelectronics technology will be more fully exploited to improve conventional concepts of computer design. This latest device has been described as a micro-mainframe, and it incorporates many features not previously employed even in mainframe machines. In particular, its architecture has been designed specifically to optimise the execution of high-level language programs, and the basic instruction set of the device is closely related to the statements of the high-level programming language ADA. The process of compiling high-level language programs into low-level machine instructions is thus greatly simplified. The new device incorporates further hardware features that greatly assist the implementation of the operating systems and memory management systems required in medium- and large-scale computers and which have traditionally been handled by complex system software. The existence of such devices will accelerate the current trend towards distributed computing systems as an attractive alternative to the traditional large, centralised computing facilities employing

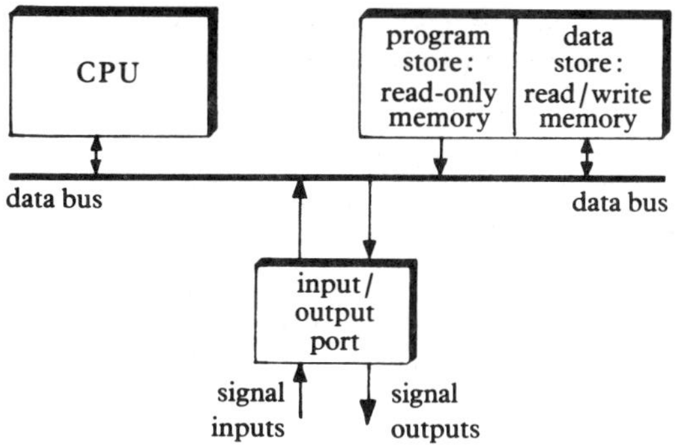

Microprocessor as a programmable logic controller. Microprocessors are often used as an alternative to conventional electronic logic circuits, to implement a specific, dedicated logic function such as the control of traffic lights. The microprocessor is required to execute only one dedicated program, which is written by the equipment designer and stored permanently in a read-only memory. The read/write memory is used only for the temporary storage of data used by the program. Input/output devices, such as operating relays to switch the traffic lights, are also dedicated to specific functions. Input and output of data in the conventional sense is not required.

mainframe machines. A distributed computing system is one in which several minicomputers, serving individual users, are interconnected in a network to allow sharing of certain resources such as large data stores.

The objective of the third- and fourth-generation microprocessors is to implement the most powerful CPU possible as a single chip, subject to the current limitations of microelectronics technology. It is clear that the advent of these devices will have a profound effect on the design of general-purpose computers and computing systems, making significant computing power available at decreasing cost to a widening spectrum of users. There are, however, many applications that do not require the sophisticated computing power offered by the third and fourth generation. In many cases a microprocessor is used, as an alternative to hard-wired logic, to implement a simple sequencing or decision-making function. The microprocessor allows the cost advantages of microelectronics technology to be obtained without incurring the high development costs of a custom chip design. Washing-machine and traffic-light sequencers are the most obvious examples of these programmable controller applications, in which the programs involved may be almost trivial and sophisticated computing capability is not necessary. The incorporation of a microprocessor into electronic equipment can have other advantages as well as cost reduction. Its capability for complex logical decision-making allows the designer of a measuring instrument, for example, to automate many functions that were previously carried out manually. These 'intelligent' instruments are sometimes described as user-friendly because they allow quite complex measurements to be carried out by relatively unskilled operators. This is of immense value in production-testing equipment, for example. Another advantage of microprocessor-based product design is that modification to the instrument, to permit optional specifications or a Mark 2 version for instance, can be more easily achieved by software changes than by the extensive re-design of a traditional hard-wired logic system.

A microprocessor that is to be incorporated into an intelligent instrument or similar equipment requires rather different features from one that is to function as the CPU of a general-purpose computer. Quite modest computing power may be sufficient and advanced features, such as memory management and operating systems, dynamically relocateable code and re-entrant procedures, are of limited value in systems in which the program is dedicated and multi-tasking is not required. It is facilities such as these that the sophisticated architectures of the third- and fourth-generation microprocessors were designed to support and these devices are not ideally suited to the programmable controller and instrumentation applications. Low cost is often the crucial factor in these systems and this requirement has been met by the single-chip microcomputer, which is the second major trend in microprocessor evolution to have emerged over the past three or four years, in parallel with the introduction of the sophisticated third- and fourth-generation CPU chips. The typical single-chip microcomputer comprises a CPU of second-generation complexity together with memory and input/output circuitry all incorporated in a single integrated circuit. Space on the chip is thus devoted to the implementation of a range of memory and other functions rather than to the provision of a very powerful CPU alone. The amount of memory is fairly small, typically consisting of sufficient ROM to store 1000 program instructions together with about 100 bytes of RAM. The devices usually incorporate a fairly large number of input/output lines and may include additional features, such as programmable timers and analogue-to-digital converters, which are commonly required in control and instrument applications. Some of these devices are available in low-power versions to meet the requirements of battery-powered instruments. Single-chip microcomputers may have excited less popular interest than the advanced architecture CPUs but, interestingly, they have been sold in considerably larger quantities.

A typical microprocessor-based instrument appears to the user as a 'black box', which has certain specific operator controls and inputs/outputs appropriate to the function of the instrument. Inside the black box is a computer, together with additional logic circuitry and, usually, some linear circuits. The computer is quite invisible to the operator, however. He cannot program it himself and he does not need to understand programming in order to use the instrument. His interaction with the computer is merely through the front panel controls, the functions of which are familiar to him and are clearly related to the applications of the equipment. The microcomputer's program, written by the equipment designer, is contained permanently in ROM and it functions, in conjunction with the equipment's hardware, to implement the required specification for the equipment. The program starts executing whenever the instrument is switched on, or when a 'reset' switch is pressed.

A typical instrument incorporates conventional electronic circuitry as well as a microcomputer, and the amount of conventional circuitry involved de-

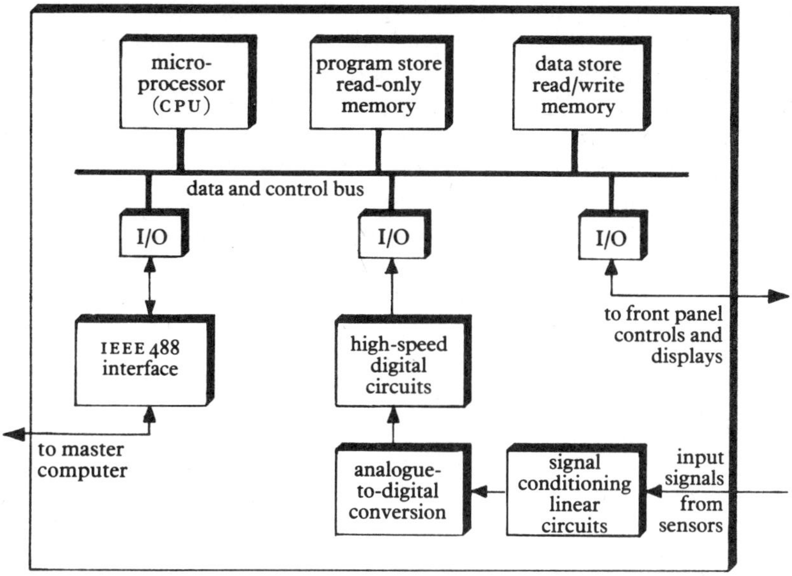

Microprocessor-controlled instrument. Inside the microprocessor-controlled 'intelligent' instrument there is a complete computer, including CPU, memory, input/output circuits and data buses, with the program permanently stored in ROM. To the user, however, the instrument is a 'black box', with front panel controls and inputs appropriate to its application. The computer controls the operation of the instrument, with linear and high-speed digital circuits being used to carry out preliminary processing of the input signal. The entire instrument may form part of a larger, computer-controlled measurement system.

pends on the particular instrument. An instrument is often required to process digital data, or analogue signals, at a rate that is too high to be handled directly by the microprocessor. It must be remembered that a computer is an essentially sequential device in which each logical operation must be carried out, in turn, by a single arithmetic/logic unit. Furthermore, the technology used to produce present microprocessor chips is one in which high speed is sacrificed to achieve circuit complexity. For these reasons, it is often found that the microcomputer is quite incapable of handling the raw data input to the instrument. Instead, linear electronics may be required to perform some initial signal conditioning on the analogue data before they are converted into digital form. High-speed digital circuitry may then be used to carry out initial preprocessing of the data, which are subsequently passed to the microcomputer in a compressed or condensed form. In general, a measuring instrument may be thought of as an interface between data and a human operator. However fast the data may be, the human operator is decidedly slow. A piece of equipment may handle input signals at a frequency of 100 MHz but the user cannot possibly assimilate information or alter the front panel controls at a rate greater than once every few seconds. The handling of front panel controls and displays is therefore ideally suited to the microcomputer and indeed, in some cases, its role is purely to control the instrument, the actual processing of data being carried out entirely by special-purpose hardware. From a cost point of view, however, it is desirable to minimise additional hardware and the choice of an appropriate trade-off between software and hardware is essential to the design of microprocessor-based instruments.

The advent of 'intelligent' instruments will have a far-reaching effect on the whole subject of measurement technology. For example, it was traditionally regarded as essential for the electrical output of a transducer, such as a strain gauge or a temperature sensor, to be linearly related to the measured physical quantity. Non-linearities caused errors in the measurement that could only be avoided by manual correction of the data using a calibration curve, a tedious, time-consuming and error-prone exercise. With a microprocessor-based instrument, however, it is easy to correct automatically for transducer non-linearities. Data defining the transducer calibration curve are stored in ROM so that when the transducer output is sensed its value can be corrected and an accurate result may be displayed.

Self-test and automatic calibration are other valuable innovations in intelligent instruments. An accurate calibration source, such as a stable reference voltage, can be built into the instrument so that every time the instrument is switched on, and once every few hours during operation, the microcomputer controller causes the calibration source to be internally switched to the input of the instrument's measuring circuitry. The value of the calibration voltage is permanently stored in ROM and the errors in the measured data can be determined. Subsequent readings of the input may then be corrected to allow for the observed drift of the instrument's circuitry. If the drift out of calibration reaches a pre-determined limit an alarm is raised, indicating that the instrument requires maintenance. These features can improve measurement accuracy and also reduce the costs of routine maintenance and servicing.

The ergonomics of instruments have also benefited from microcomputer control. Modern instruments often employ a single keyboard and display in place of the confusing array of switches and knobs employed in the past. The setting-up of a complex instrument can be greatly simplified by using an interactive procedure in which the microcomputer asks the user, by means of the front panel display, to enter desired parameters using the keyboard. The value entered may then be repeated on the display, giving an immediate check. Complications such as coarse and fine controls can be eliminated, their function being taken over by appropriate arithmetic operations within the microcomputer. The selected parameters can be continuously displayed for immediate reference. The voltage sensitivity and time base of an oscilloscope, for example, may appear as figures at the top and bottom of the display screen. The front panel keyboard is one of the few ways in which the computerised nature of the instrument may be apparent to the user.

Another feature of modern instruments is the ease with which they may be incorporated into larger, computer-controlled, automatic measurement systems. A medium-sized minicomputer may be used to control the operation of several individual instruments, according to a pre-determined program. The instruments usually include stimulus devices, such as power supplies and signal generators, as well as measurement devices, such as voltmeters. The central computer sends commands to the stimulus devices to define their outputs (for example the frequency and amplitude of a sine-wave generator). It then instructs the measurement devices to monitor the performance of the system under test, in response to the specified stimulus. The stimulus can then be changed and further measurements made so that a complete test procedure can be built up. This kind of system has been greatly facilitated by the definition of standard procedures for data communication, notably the IEEE-488 instrumentation interface specification. The implementation of data communication interfaces in individual instruments is greatly simplified by their computer-based nature.

It is clear that the microprocessor has promoted far-reaching improvements in the design of electronic instruments and systems, as well as in the field of general-purpose computer systems. The development of microprocessor-based equipment naturally has its pitfalls, however, and it is not surprising that many expensive mistakes have been made by manufacturers in attempting to introduce microprocessors into equipment design. The most severe problems have arisen from the under-estimation of the cost of writing software, due perhaps to the intangible and abstract nature of this activity which, by its very nature, does not result in a visible, physical product. The development of software involves much more than merely writing the code. The completed program must be exhaustively tested to discover the inevitable logical errors (bugs) contained in it and these errors must then be corrected. Proper documentation of the software is vital to allow subsequent maintenance and modification of the product. Productivity in software development can be greatly improved by the use of programming aids such as high-level languages, which have been well established in the computing industry for many years.

Capital investment in equipment is also necessary for the efficient design and development of any but the simplest microprocessor-based systems. The traditional oscilloscope, or even logic analyser, is an inadequate tool for the development of a system that involves parallel data (sixteen-bit addresses and eight-bit data words) and complex logic structures such as repetitive program loops and subroutines. Special microprocessor development systems are available that provide facilities such as assemblers, compilers and simulators to aid the development and testing of software. They also incorporate facilities that can assist in identifying the particularly troublesome faults that may arise when software is required to function in conjunction with newly developed hardware. Development systems are expensive, however, typically costing $40,000, and newcomers to the field are often surprised to find that such a large investment is needed to support the development of systems based on a microprocessor chip that itself costs perhaps $10 or $20.

The rapid development of microprocessor technology has itself posed problems for users. The availability of a new, more advanced type of microprocessor requires the equipment designer to devote considerable time to familiarising himself with the operation of the new device before he can employ it

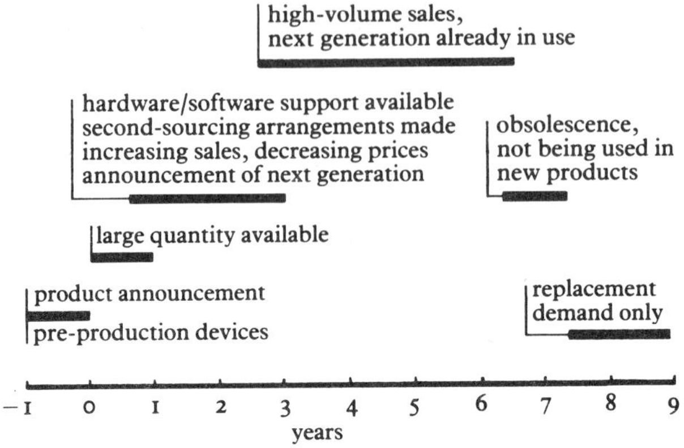

Life cycle of the microprocessor. The life span of the microprocessor, from its initial announcement to obsolescence, is about 8–10 years. It may be three or four years after the device announcement before it begins to be widely incorporated into equipment, and the next generation of more advanced processors may well have been announced by this time. The original device will not become obsolescent until its successors have reached the 'mature' stages in their own life cycles. (The time scales are only an approximate indication of typical trends.)

effectively in his products. It may also be found that the new type of processor is not supported by an existing development system, in spite of claims about the universal nature of this equipment. Furthermore, the equipment manufacturer may be reluctant to base his products on a new device that is obtainable from only one semiconductor company. The application of advanced microprocessor technology has, as a result, been rather slower than the progress in device technology itself. The first- and second-generation microprocessors have had life-cycles of around eight to ten years, and a substantial time-lag exists between the introduction of a new device and its impact on equipment manufacture. It is a consequence of the highly competitive nature of the semiconductor industry that a new microprocessor design is usually announced, and aggressively marketed, well in advance of its availability and perhaps even before its design is completed. The equipment manufacturer who proposes to incorporate the device into a new design of instrument must therefore be aware of the effect on his own product development programme of delays in the availability of devices. Even after sample quantities of the new chip are available, it may be a year or more before it can be obtained in production quantities. The availability of hardware and software support for the device is also essential to the equipment user. Evaluation kits, single-board computers based on the new processor, assembler and compiler software, and software to allow existing development systems to be upgraded to support the new device, are all required to support the cost-effective development of microprocessor-based equipment. Second-sourcing arrangements, whereby identical components can be made available from more than one semiconductor manufacturer, must also be established in order to give customers confidence in the reliable and continued availability of the new device.

With the establishment of second-sourcing agreements and the general availability of hardware and software support, a new device will begin to be widely incorporated into equipment designs. As this equipment itself enters large-scale production, demand for the chips increases and their price falls rapidly as the semiconductor companies recover the capital investment that they made to develop the new device. The lower price promotes wider use of the device and a period of very high volume sales begins. Bearing in mind that new generations of microprocessor have been introduced every four years or so, it can be expected that the initial announcement of the next generation of devices may well be made even before the previous generation has begun to sell in large volumes. From a purely technological point of view, it is probably true to say that microprocessors have often become obsolete before they have been used. We have seen however, that the commercial viability of a device depends on much more than the availability of the chips themselves and it is only when the succeeding generation of devices itself reaches maturity, with high-volume production, that its predecessor may be regarded as entering the final, obsolescent, phase of its life-cycle. Its use in new equipment then declines, although the devices will continue to be produced for many more years, in response to the demand for replacements in existing equipment.

It is a matter for concern that the terms 'microprocessors' and 'microelectronics' are apparently not clearly distinguished in the minds of many laymen, administrators and politicians, and the frequent use of the composite description microelectronics/microprocessors is indicative of this serious misunderstanding. The microprocessor is only one of a wide spectrum of microelectronic devices. It was not a revolutionary invention but, rather, it was a natural step in the evolution of microelectronics technology, which had its origins thirty or more years ago. Nor should the microprocessor be regarded as a revolutionary development in computer technology. Indeed, the primitive architectures of the early microprocessors and the associated use of crude, low-level software methods have prompted the comment that the microprocessor 'set computer science back by twenty years'. The microprocessor represented a breakthrough only in respect of the dramatic reduction that it made possible in the cost of sophisticated information-processing capability. Small, single-chip, logic-replacement microprocessors have brought the cost advantages of LSI technology to a wide range of electronic equipment. The third- and fourth-generation advanced architecture devices have opened a new market for extremely low-cost, yet sophisticated, general-purpose computers while bit-slice microprocessors have greatly reduced the cost of conventional, high-speed minicomputers. The microprocessor represents a cost-effective solution to many design problems at the present time but it must not be forgotten that rapid developments are taking place in other aspects of microelectronics. Improved techniques for the computer-aided design of integrated circuits may soon reduce the cost of designing special-purpose chips. Semi-customised devices such as programmed logic arrays (PLA) and uncommitted logic arrays (ULA) are also becoming increasingly attractive. Furthermore, it is confidently expected that continuing improvements in device technology will lead to a further order of magnitude increase in chip complexity in the near future. This will bring additional problems of effectively exploiting such complexity, and the microprocessor as we know it today may well be succeeded in the future by alternative kinds of customised or programmable microelectronic devices.

The achievements of the semiconductor industry in producing increasingly complex chips at very low prices are clearly impressive. It must be borne in mind, however, that microprocessor-based products involve much more than the microprocessor chip itself. A computer requires peripheral devices such as visual display units (VDU) and printers to which it must be interfaced. On-line computers (such as might be used in the automation of a factory production line) are usually required to interface to analogue data (for example to measure a voltage) and to control electro-mechanical devices such as motors and solenoids. A great deal of electronic circuitry (hardware) is therefore required in addition to the microprocessor and memory devices themselves. As well as the hardware, a microprocessor-based product requires software – the program that causes the computer to operate in the required manner in a given application. The same microprocessor chip may be used in a wide variety of different applications and it is the software that determines the particular characteristics of each system. The microprocessor industry therefore has two distinct facets. The first is the manufacture of the microprocessor chips, a task that is generally carried out by semiconductor companies, and the second is the design and construction of microprocessor-based systems or equipment, which may be described as microprocessor applications engineering. It is important to understand that these two activities are quite different in nature. They are based on different scientific disciplines, require different skills in their workforces (both technical and manual) and they involve very different levels of capital investment and different types of market.

The advent of the microprocessor is clearly having a profound influence on the design of electronic products and computing equipment. The disciplines of electronic engineering and computer science are converging and it is essential that the designers of microprocessor-based equipment develop a professional approach to software development. It must not be thought, however, that electronic equipment design has been reduced to a matter of mere programming. The designer of a measuring instrument, for example, must have a thorough understanding of the physical parameters involved and of the human environment in which the instrument is to function. These factors will determine a system specification. Its cost-effective implementation will generally require a judicious blend of hardware and software techniques, and microprocessors are only one of a wide range of components that the engineer will require to use in order to arrive at an effective solution. The microprocessor represents a powerful and exciting addition to the electronic engineer's tool-kit but it is not the best solution to all problems. Cost-effective products will only be designed by engineers who can correctly assess the relative merits of microprocessors and other methods and who adhere to sound principles of engineering design, which apply equally to the development of both hardware and software.

4. HOME AND OFFICE COMPUTERS

The appearance of computers in many, diverse application areas – from the High Street automatic cash dispensing facility, to the microprocessor-controlled washing machine – has been made possible by new microelectronic technology. These mass-produced chips are presently available for the price of a paperback book and computers with very respectable processing powers (comparable to major commercial installations in the sixties) are now cheap enough for the home enthusiast. These inexpensive systems have also found their way into the office, where they are used to carry out accountancy and general book-keeping tasks. One of the most useful of these applications has been the introduction of the word processor, a computer program that enables the user to prepare manuscripts in a much more flexible way than ever before. Here the main incentive for the computerisation of office tasks must surely be the reduction of paper flow where, in any medium-sized organisation, about 85 per cent of paperwork is concerned with information exchange between departments. If access to a computer terminal were freely available to all members of the organisation, much of this paper wastage could be dispensed with.

Some of the apprehension about the computer invasion has arisen because the majority of people do not understand how computers work, and because they have no conception of their ultimate limitations. It is therefore relevant to consider how computers function in more detail, before going on to consider what computers can do at present and what they may be able to do in the future. Computers come in two basic forms, analogue and digital. Nearly all modern computers, unless they are made for a very specialised application, are digital computers. For this reason most modern texts refer to computers as meaning digital computers, the adjective digital only being found in the titles of texts more than ten years old. An analogue quantity is one which can vary continuously, like the temperature of a room, whereas a digital quantity is one which is limited to a set of discrete values, like an on-off light switch. In computing, the digital quantity is normally voltage and a digital computer is based on the binary number system where one voltage (often +5 volts) is used to represent a binary 1 and another voltage (usually zero volts) is used to represent the binary 0. Thus streams of 1s and 0s can be transferred from place to place within a computer, along a single wire. It is usual to have several wires in parallel in order that more information may be passed in a given time. When this happens the group of wires carrying the digital information is termed a bus. Commonly, groups numbering 8, 16 or 32 are used. The basic unit of binary information is the bit, and so a bus of eight lines will have the capacity to carry eight bits of information in parallel – a byte. Therefore, a bus of sixteen-bit capacity can carry two bytes of information in parallel.

A typical computer system consists of three main parts, the central proces-

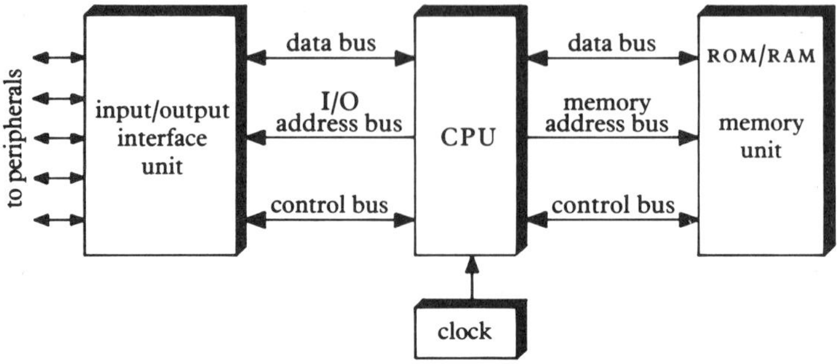

Computer structure. The main components of any computer are the central processing unit (CPU), the memory, the input/output (I/O) unit, and the buses. The CPU performs the arithmetic and logic functions of the computer, as well as the required control functions. The memory includes both read-only memory (ROM), to hold the machine instruction set and other fixed data, and read/write memory (RAM), which stores the data being processed. The I/O interface unit contains control logic to manage the transfer of serial or parallel data streams to the various computer peripherals such as printers, displays and keyboards. There are at least three buses linking the main computer components: the address bus, to indicate required memory locations; the data bus, for transfer of data within the computer; and the control bus, which determines which operations are carried out.

sing unit, the memory, and the input/output circuitry, all of which are connected by buses. Large-scale integration of components has resulted in the central processing unit (CPU) now being available as a single chip. There are normally two buses: the data bus transfers information and the address bus determines the location in the system to which this information is to be sent. A clock is necessary in order that all system operations may be properly synchronised. The input/output circuitry is the computer's only means of communicating with the outside world, and external input/output devices can include printers and video display units (VDU). The memory unit contains all the programs and data necessary to provide the user with the facilities he desires. Memory comes in many varieties and may be broadly categorised by its speed, that is, by the time it takes to respond to a request from the CPU. In general, the slower it is the cheaper the memory, and this fact has led to the use of programs, known as operating systems, which store the bulk of the data and programs in the slow memory and organise the transfer of current material into a smaller amount of fast memory. The operating system gives a computer its personality, since it is this program that has a repertoire of commands enabling it to co-operate with the user. A good operating system will hide the internal structure of the system by carrying out all the various memory management tasks and scheduling of input and output in an autonomous manner.

The central processing unit is the heart of the computer and is able to interpret and act upon instructions that have been stored in memory. It contains a large amount of control logic and also a set of registers for temporary work space. These registers are a form of memory, but because they are fabricated on the device, they have been optimised for speed of access. The CPU understands a very low-level language, the function of which is intimately bound up with operations that are executed on these registers. As an example of this, every computer has the ability to move data from external memory, along the data bus and into one of the registers. Certain of the registers can have data from external memory added to, or subtracted from them. There will always be a register known as a program counter, which keeps track of the next instruction to be executed. This is done by keeping a note of the address in external memory where the next instruction is stored so that, once this instruction has been fetched into the CPU, the address in the program counter register is incremented and it addresses the next instruction, and so on. It can be seen that the size of the program counter register is determined by the number of lines in the address bus. If the address bus has sixteen lines then the program counter register must be able to store sixteen bits of information. A sixteen-line address bus gives the possibility of addressing up to 65,536 locations.

In the same way, the amount of information that can be stored in each external memory location depends upon the number of lines comprising the data bus. If the data bus is eight bits long then each unit of memory is likely to be the same length. It is, however, quite feasible to organise the memory in units of sixteen bits so that two groups of eight bits could be sent along the bus in sequence. These would then be stored side by side in the double-length memory location, using some extra control logic. The length of memory unit used in a computer is known as the word length for that computer and the word length can vary depending on the type of computer.

The fastest type of memory in current use for small computers in the home and office is semiconductor memory and the bulk of this will be read/write memory (RAM). Semiconductor RAM is, in general, volatile; that is, once the system is switched off, any information that was contained in memory is lost. For this reason most small computer systems also have a small amount of read-only memory (ROM), which usually contains sets of utility routines and initialisation procedures that enable the system to start up properly when it is first switched on, and will have been written into the ROM in the factory. A note on terminology is relevant here. Everyone is probably familiar with the term hardware. This usually refers collectively to all the electronic components that go towards the realisation of a computer system. Widespread use of this word has led to the introduction of the term software, which refers to the collection of programs and data that reside on a particular computer system. The choice of the word soft is intended to convey the idea that programs are not as permanent as the hardware, i.e. they may be easily altered without changing the hardware in any way. When devices such as read-only memories are considered this is not really true, and so the term firmware has been coined to described the programs that reside on these devices.

There are three common types of slow memory: hard disc, floppy disc and magnetic tape. A hard disc is in appearance a little like the more familiar discs used for sound recording purposes, but instead of a groove that spirals inwards from the edge to the centre the disc memory consists of a set of concentric rings of magnetic material. The disc is read from and written to by a magnetic head, which can move back and forth across the surface. The floppy disc is similar in principle to the hard disk, but it is much cheaper and, as the name suggests, it is flexible. Magnetic tape has much slower access times because the information is stored serially along its length, and information retrieval involves winding the tape back and forth to find the specific position required on the tape to gain access to the information. Thus, although it can be used to store large amounts of data, it is only used for the storage of information which is seldom used. A few more exotic forms of storage media have appeared recently as sub-system components of new products, but these will be mentioned later.

The most common means of inputting data into a computer system is via a typewriter-like keyboard. These keyboards often form part of a small terminal system, which has a television screen attached, allowing the user to display the information currently being typed in. The display screen is also used to examine programs and data that have previously been stored in the system. Any complete system also requires some form of hard-copy machine, possibly by having the keyboard form part of a terminal that interfaces to a printer instead of a television screen (rather like an electric typewriter). Such devices are quite common, but they tend to be slow and they waste large amounts of paper. It is often better if the text is created and edited using the visual display screen and then printed out as a finished product. For this reason, many stand-alone printers are available, and these may be connected (via a suitable interface device) to the system bus.

The type of print head used will determine the quality of the output. The most common is the dot-matrix head, which consists of a small matrix of pins programmed to configure into the desired printing character as each one is selected. For high-quality printing a daisy-wheel head, which has a central hub and thirty or forty spokes with raised mirror images of the characters on the ends, can be used. Characters are selected by spinning the wheel round at high speed until the desired character is in line with the striking hammer. Another advantage of this type of printer is that the daisy wheel is easily changed should a different font be desired. Many applications require more sophisticated output methods for the display of drawings and graphs. To satisfy this need there is available a type of visual display unit known as a graphics terminal, which provides, at a somewhat greater cost and possibly in colour, the extra resolution and flexibility to give the required graphics performance. Graphics plotters serve as the hard-copy medium for graphics output.

The best example of computer impact on some of the more conventional office methods is the word processor. The first step in the preparation of a report traditionally involves draft being passed to the typist. This is typed and passed back to the author for inspection. The author will indicate any typing errors and other corrections to the typist. The correction process will be done in a number of cycles, each involving painting out errors and retyping over them. If more than a word or two of extra text has to be added, the remainder of a page may have to be retyped.

After gaining familiarity with the use of a computer to aid document preparation, this conventional method seems extremely inefficient and time wasting. The draft text may be entered onto computer file, either by a typist, or by the author. Short formatting directives are entered along with the text and these serve to mark out new paragraphs, headings, etc., for later processing by the computer. The author can then check for errors and correct them with the aid of the text editor, a program which enables the user to modify text: replacing, deleting, or adding whole sections if necessary. When the text has reached a satisfactory state of development, the file is run through with another program, which processes the formatting directives and fills out lines to give the required format. Pages and sub-section headings can also be numbered if this is desired. Text processors come in packages with special programs that are able to carry out further valuable tasks such as counting the number of words in a file, or even checking for spelling and typing errors. Yet another advantage of storing documents on computer file is that they may be easily modified or updated at a later date. This ability to manipulate information files is probably the main feature that distinguishes the computer from the programmable calculator.

In addition to document preparation, one of the most powerful uses of modern computers is in database storage and retrieval systems used to maintain data on almost any subject known to mankind. For instance, a large company may wish to keep a stock inventory of all its products, organised in such a way that items of a similar type are grouped together and may be accessed by a single key name. Subsections are then found by further keys as the subject area is localised to the one of interest. A great deal of work has been done in an effort to produce a natural language interface to these systems so that an inexperienced user can still find the information he requires. English, German and French are natural languages, but because they have developed over thousands of years their precise grammar is extremely complex and the task of designing a computer program that can parse any sentence successfully has proved to be insuperable up to the present time. Parsers have been written to command a very respectable subset of the English language, and whenever a sentence is not understood by the parser it can send the user a message asking for the sentence to be rephrased. This approach is fairly effective but many ambiguities can occur, which only humans have the power to resolve. Such database management systems, which are available with or without a natural language interface, are becoming an extremely important part of business life and possibly the most profound effect that modern computer technology will have on society will be through the rapid dissemination of information through database systems.

It is interesting to observe the publicity that has been given to the emergence of an inexpensive ($100) pocket television. This device, which is marketed by Sinclair of Cambridge Ltd, seems to have caught the eyes of the media much more effectively than a similarly priced computer introduced by the same company. Although the Sinclair ZX80 computer is selling well (about 50,000 were sold in its first year), the sales of the pocket television will probably be counted in hundreds of thousands and possibly even in millions. Even the later

Personal computers. The range of applications to be found for the home or hobby computer are ultimately limited only by the inventiveness of the user. Various software (program) packages are commercially available to perform such tasks as music composition, housekeeping accounts and educational games. In addition, the home computer can also be used to 'control' the house in terms of burglar alarms and central heating systems. *Photograph courtesy of Microsense Computers Ltd.*

announcement that a cheaper version of this computer (the ZX81) would soon be available had little impact. This tends to suggest that, given the choice, 90 per cent of the population would buy a new type of television before they would consider buying a computer. There also lingers the suspicion that, like the pocket calculator, the cheap home computer will in many cases spend most of its life on a shelf gathering dust. This is because learning to use the present generation of small computers requires the kind of effort one would normally need to learn to play a musical instrument. The effect of this barrier has been to limit the home computer to the educational and hobby markets only. The present state of home computing as a minority activity will probably change as more intelligent computer interfaces are produced for the consumer market.

Computer games, on the other hand, are becoming fairly commonplace around the home. Again they seem to suffer the same fate as many pocket calculators, that is, they are discarded as soon as the initial euphoria of having a new toy wears off. Even if the keen player perseveres past the novelty stage, most of the currently available games are soon beaten, or at least exhausted of interest. The main exception to this unfortunate fact of life lies in the area of computer chess. Many different products are available at quite reasonable prices and they do play to a very high standard. The reason for the high quality of chess playing programs is, perhaps, not obvious to the newcomer on the computer scene and so a word or two of explanation is in order. When the first modern digital computers became available in the early sixties, many research workers in the then infant field of computer science felt that chess would be a suitable model for the study of computer-based problem solving. It was thought that the experience gained would lead to the discovery of problem-solving strategies that could then be applied to more general real world models. This approach was only partially successful because, although new insights were gained which were later applied to other areas, no general framework was found. Notwithstanding this failure, for the last twenty years a great deal of effort has gone into the study of computer chess and one of the benefits of all this work has clearly been the development of highly efficient chess-playing programs. That the work has not proved to be more generally applicable is also clear, from the poor standard of other computer games.

In a different category completely are the Space Invader type games, where the provision of a graphics display enables the manufacturer of these systems to take advantage of one of the principal strengths of computers, namely, speed. The development of games that challenge the speed and accuracy of the user has now produced an interesting future in that the complexity and graphics impact of games require to be continually improved as the system programmers try desperately to keep one step ahead of the enthusiasts. The computer, in the guise of the microprocessor chip, will invade our homes in many other ways. Microprocessors are already being designed into common household equipment such as cookers and washing machines and, provided they are not too demanding in terms of operator skills, their use will soon become much more widespread. One very important future development may be the use of a microprocessor system to monitor heat loss in the home.

The use of the computer as an aid to the performance of general office tasks has recently been made possible by the development of cheap personal computers with enough power to execute programs that have, in the past, been the domain of the mainframe computer. Programs that can carry out book-keeping, accountancy, word-processing and even simple database management are available for personal computer systems manufactured by firms like Cromenco, Sharp, Apple and Commodore. Most of these systems are based on eight-bit microprocessors like the Z80. Operating-system software and many other useful programs usually come with the system, but much more is available through software houses that specialise in producing programs for specific systems. Systems based on other microprocessors, like the Motorola 6800 and the Rockwell 6502, are also available although the software availability is more limited. In these cases, their use has become much more widespread in the USA, and so software is obtainable through UK-based subsidiaries of American companies. Care must be taken in the choice of software since compatibility between systems is not always good. Programs may be written in either assembly language or a suitable high-level language. An assembly language is characterised by a set of mnemonics, which the programmer may use to ease the task of writing the low-level machine-code that is

Office computer system. Small, inexpensive and easy-to-use microcomputer systems are now available to suit the needs of the smallest of companies. Facilities for word processing and document preparation can be incorporated, using the keyboard and visual display unit to create the text, the central processing unit and floppy discs to store the information, and a high-quality printer to produce the final document or letter. *Photograph courtesy of Microsense Computers Ltd.*

understood by the internal logic of the central processing unit within the microprocessor itself. The mnemonic program listing is then processed by another program, the assembler, which translates the mnemonics into machine readable code that may then be executed by the microprocessor. For example, the instruction 'LDA 3' means load accumulator (A) with the contents of memory location 3. The output of a Motorola 6800 assembler, when given this as an input, will be '10010110 00000011'.

A high-level language, on the other hand, is little influenced by the internal structure of the computer being used, so that the same program can be run on any system providing that it supports a suitable compiler. A compiler is a program that translates the high-level language into machine-code ready for direct execution on a specific computer. Thus a separate compiler has to be written for each microprocessor type before a high-level program is transportable between one system and another. This can sometimes lead to small incompatibilities between one system and another, since a given compiler is often written by a firm allied to the production of one particular type of microprocessor. Therefore, the interpretation of the language definition will vary slightly from one system to another and these slight variations can lead to time having to be spent in converting a program for one system to run on another system. However, this effort is always very much less than having to rewrite the program at the assembly code level. Another advantage of high-level programming is that these languages were specially designed for clarity and ease of implementation (although, to the inexperienced, this sometimes seems hard to believe).

For a given software applications package (on floppy disc or magnetic tape) the program will often be in object code format. This is the machine code listing as produced by a compiler or an assembler. It is important that this package should be matched precisely to the system as there is no easy way of modifying it should it not be completely compatible. An alternative approach here is to use the source code listing. The source code is the listing of the program in whichever high-level language it was originally written. This is usually more expensive and also means that a suitable compiler must be used to process it. Nevertheless, possession of a compiler for a popular high-level language means that, with a little effort, most programs may be modified to run on most systems, even if they are not directly compatible. Yet another advantage of using high-level languages is that they do not depend upon the word size of the system on which they reside. For instance, a program on an eight-bit microprocessor system is essentially the same as the program on a minicomputer, or even a mainframe. This is important if at some later date the system hardware is upgraded. In fact, this question is very pertinent to the problem of which scale of computer system is selected for a specific purpose. Until recently, computer systems were cleanly partitioned into three size and performance groups. At the small end, there was the eight-bit microprocessor-based personal computer. The next stage up was the minicomputer (around sixteen-bit word size), the most popular of these being the DEC PDP11 series produced by Digital Equipment Corporation. Next up the scale came the mainframe, the IBM 370 system being a typical example of this. The cost and performance increase associated with each move up the range was about ten. The fast moving pace of microelectronics, however, has once again altered the situation dramatically. The mainframe manufacturers have brought out bigger and faster systems and DEC have produced a system that goes more than halfway between the mini and the mainframe (these have been termed midi-computers). An example of this type of system is the DEC VAX780. The most exciting recent development, however, is the advent of the sixteen-bit microprocessor. This device has the power of a good minicomputer and because it is fabricated on a single chip, its price will soon bring sixteen-bit microcomputers into the personal computer market. The next effect of these developments is that instead of having only three groupings of computer types, there is now an almost continuous range of product powers from the Sinclair ZX81 right up to an IBM 3081 mainframe.

The development of the sixteen-bit microprocessor will have very important consequences in office computing. It is therefore worth considering how this area is most likely to develop. The best approach is to take a look at the design of operation systems, since there are three possible approaches for the sixteen-bit machine. The first is to develop an entirely new operating system, hopefully avoiding some of the weaknesses of current operating systems. The second

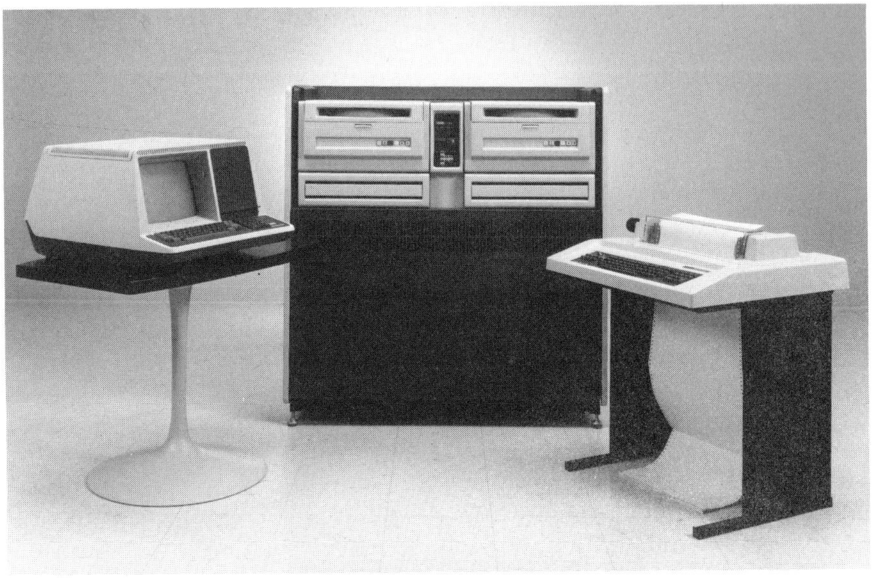

Minicomputer system. Minicomputers require a minimal level of operating staff and are consequently ideally suited for general-purpose use by medium to small companies and offices. Time-shared operating systems are available (UNIX) to allow several users to access the minicomputer simultaneously, running programs for financial control, stock control, invoicing, word processing and (personnel) records. A typical system as shown here (PDP 11/60) incorporates two disc drives, operator console and line printer. *Photograph courtesy of Digital Equipment Ltd.*

approach is to upgrade an eight-bit operating system and the third approach, the one most likely to succeed, is to modify an existing minicomputer system. The system that appears to be preferred at the moment is known as UNIX, which was originally developed at Bell Telephone Laboratories, and seems to have already made its mark as a very successful operating system using PDP11S. Microsoft (a software house) is already advertising an operating system known as XENIX, a customised version of UNIX, which is available for the Intel 8086, the Zilog Z8000 and Motorola's MC6800. Other UNIX-like systems in the field are ONYX, Cromix (Cromenco), UniFLEX and OS-1. Although the operating system is of little interest to the computer user, selection of a computer with a powerful operating system such as the UNIX will enhance the useful life of the system.

Another factor that is important to the choice of computer system is whether or not to run a time-shared system. That is, will the computer system need to respond to more than one operator at a time? Time-shared systems have many VDU terminals connected to the same computer and the computer divides its processing time amongst the currently active users. If the computer is fast enough (and the operating system good enough) each user should feel that he has a computer system which is dedicated to his use only. Of course, if too

Mainframe computer complex. The complex shown here (IBM 3081) is a good example of a modern mainframe computer system. Systems of this type are general-purpose machines and can be found in manufacturing industry as well as in banking, insurance and government departments. A more specific example of its application is as a central database in airline flight reservation systems. The system has several excellent features, such as colour display terminals, printers and monitors, in addition to its extremely large memory capability and fast processing speed. *Photograph courtesy of IBM United Kingdom Ltd.*

many users are on the system at the same time the system response becomes very slow, so the machine power required is determined by the number of potential users and by the type of computations being carried out. The largest consumers of processing power are organisations that require large amounts of numerical computation. Fortunately, this is a task which is unlikely to be high on the requirements list for office-work. If the requirement is for more than two regular users of the system, then the eight-bit personal micro-computer is not sufficiently powerful since these systems are not really fast enough to run a satisfactory time-shared system. Because of this, most eight-bit-based operating systems are designed to be single user only. Incidentally, this is the main reason why upgrading eight-bit operating systems to sixteen-bit is going to involve more work than adapting operating systems from a minicomputer. If a time-sharing system is required, a system based on a sixteen-bit microprocessor should be considered, because such systems will be much cheaper and almost as powerful as the current generation of minicomputers, and recent advances in mass storage systems mean that these systems are becoming even more attractive in terms of price to performance ratio.

Although the floppy disc has found successful application in the personal computing market, it is not reliable enough, fast enough, or rugged enough for timesharing use. Probably the best configuration here is a small hard disc cartridge, the most notable of which is the Winchester (and known as the Mini-Wini). Some companies are offering systems with one small cartridge disc for general work and a streamer for general system management tasks. A streamer is a new addition to the mass storage repertoire. It is a magnetic tape that is not on a reel, but is bunched up into a cartridge and can therefore be moved past the magnetic heads at much higher speeds than conventional tape drives. The streamer can be used to perform backups, i.e. taking regular copies of the operating system disc in case of a system crash. If the disc has become corrupted, the state of the system at the last backup can then be restored.

Another consideration affecting the selection of an office computing system is whether or not the system should be connected to another site, to ease the problem of having geographically distributed offices. If the distance between offices is too great for a direct connection, the telephone network may be used to complete the link. The telephone network may also be used to connect an office with a remote database should access to such a system be necessary. At present the telephone network is an expensive solution, although cheap rates are available for passing data between systems during the night. Data-communication systems are a very active area of research and future systems based on packet-switched networks will soon alter the situation yet again. One of the major problems of having distributed sites is the problem of keeping consistency between the various sites. Any database system with distributed copies must ensure that inconsistencies do not arise should one of the users at a distributed site modify that copy without informing everyone else of his actions. This is one reason for using cheap rates overnight so that daily updates at the various sites can be broadcast to all the other sites to achieve overall consistency before the start of each working day. However, this is a simplification of the real situation and much work still has to be done in this area.

What then of the office of the future – an office with no typewriters, no filing cabinets and, instead of a desk, a computer console? The biggest setback to the realisation of the electronic office is the problem of how to move to that state from our present state. Filing cabinets, etc., will have to remain in all offices as long as a few companies still stick to conventional methods and the most likely scenario is that there will be a gradual increase in the number of offices that support computer facilities. This will go on until a position is reached where use of computers is sufficiently widespread to permit offices to carry out business only with other companies that can support the same facilities as themselves. When this happens, the less efficient companies which have held on grimly to the older methods will rapidly disappear. Unfortunately, this growth mechanism will lead to offices becoming computerised in a piecemeal fashion, creating a position rather like the railways in the nineteenth century, where different companies operated on lines with different gauges. However, like the railway system, at some stage the pressures will become great enough to force everyone to start using equipment that conforms to some universal compatibility standards.

One of the greatest inefficiencies in even the best electronic offices of today lies in the present method of entering text into the computer keyboard, since this restricts the rate of data entry to that of the fastest typist at best. The obvious solution is to build a speech recognition system into the front end of the computer so that data can be fed in at rates comparable to human speech, enabling even the most inexperienced computer user to enter text quickly. In recognition of this fact, a great deal of research is currently being carried out in an effort to realise this possibility. Predictions are being made that integrated office systems will appear within the next five years. However, users have to accept office systems before they will take advantage of them. Therefore, such systems should be designed to minimise learning effort, because the current lack of awareness and cost of retraining will be the real bottlenecks in the path of future progress. Much more attention has to be directed towards making computer interfaces user-friendly

What about new technology? Are more surprises just over the horizon? As far as memory units are concerned, many more new devices are waiting in the wings. Philips have developed an optical recording system that can store a million megabytes of data, with access times of less than a second. Bubble memories, a method of storing information in magnetic domains, have still to make an impact as an integral part of small computer systems. Facsimile images of ordinary written material or drawings can be transmitted over telephone lines, so that electronic mail will eventually become an accepted part of the electronic office. User terminals with built-in microcomputers and communications modems are now being designed to fit inside a small suitcase. Such devices could serve as portable workstations and, assuming the existence of a good communications network, these workstations could be plugged in wherever the user happened to be. This tends to suggest that the integrated electronic office concept is a fallacy and when microelectronics achieves its ultimate level of impact, the office need no longer exist. A future rewrite of this chapter may be called just 'Home Computers'!

5. MICROELECTRONICS IN CONSUMER PRODUCTS

The microelectronics revolution has had far-reaching effects on the electronics consumer market, which, although not a new market, was previously precluded from developing into a mass market by the fact that available electronics products were too expensive. This has changed with the advent of cheap, reliable microelectronics and the emerging ranges of consumer products utilising microelectronics seem boundless. Microelectronics and microprocessors are no longer used solely for specialised scientific equipment but are appearing in such diverse products as toys, games and domestic products. Even the motor car, which is predominantly a mechanical product, has been affected and it is now possible to fit microprocessor systems in the car to monitor several functions within the engine and to provide the motorist with up-to-the-minute trip details such as estimated time of arrival, fuel consumption and time of day.

To most people, the television receiver represents the single most widely accepted consumer product based on microelectronics. Throughout the years, improvements in television designs have rapidly exploited all the developments in microelectronics up to the present day where the microprocessor looks poised to influence television design. The early integrated circuits were used to substitute less reliable and often more expensive components. For example, mechanical parts such as tuning dials and push-button switches were replaced by integrated circuits for preset tuning and for touch-sensitive switches, a phase of television development known as the substitution phase. The introduction of colour television, in itself made possible by the availability of new microelectronic integrated circuits, led to a phase of innovation in television design. For example, electronic display of channel numbers and electronic programme tuning became available. Possibly the most successful innovation here was the development of the remote control handset for channel selection, volume and contrast control. Here either an inaudible tone (ultrasonic) or an invisible beam (infra-red) is sent from the hand-held transmitter to the control receiver fitted to the television. Integrated circuits within the TV set decide the appropriate actions required to comply with the remote control instructions and adjust the television accordingly. This phase of 'innovation' is ending and current trends are towards the development of microelectronics-based peripherals and add-ons to the television receiver, using the television simply as a data viewing and information display medium. The colour television has become a natural candidate for the application of microelectronics and as a result the television itself has become more versatile, being used for a whole range of new purposes such as video recorders and video discs, teletext and viewdata information services, and extending from these towards the creation of an integrated home terminal offering recreation, information and housekeeping services.

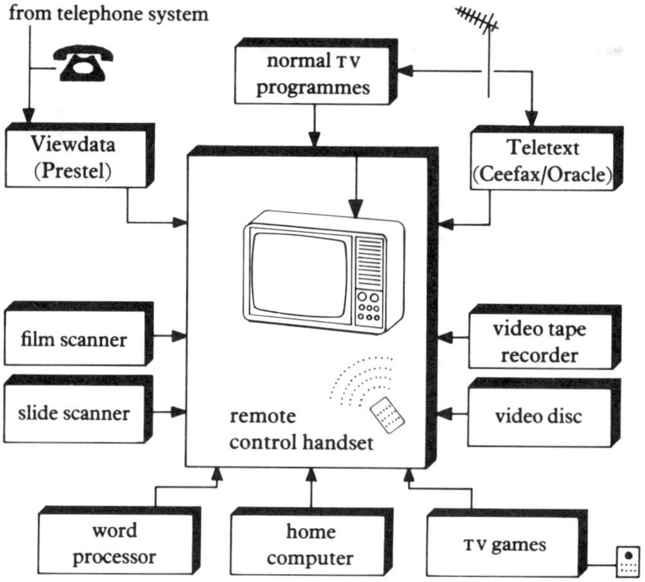

Video consumer products. The standard television set forms the heart of the home information and entertainment centre. In addition to its use for display of text and graphics from teletext and viewdata services, the television forms an integral part of video disc and video tape recorder systems offering home video entertainment, and it is predicted that the consumer market for pre-recorded video tapes (films, TV shows, etc.) will experience a dramatic growth in the eighties. A wide range of self-contained video games exists, and the use of the television with a home computer system offers scope for unlimited games applications as well as general, more serious use of the computer for housekeeping duties.

The television receiver itself is undergoing many significant changes as a result of microelectronics, some of these being apparent to the viewer and others less so. For example there has been a steady decrease in the number of components and connections used in the television, making servicing more efficient and less expensive. New, high-precision in-line picture tubes with automatically aligning deflection units remove the requirement for adjustment of features such as focus, colour quality and picture orientation. Reduced spot size and improved uniformity of focus across the screen as well as improvements in picture contrast, even at high brightness, drastically improve picture quality. New technologies, such as surface acoustic wave (SAW) technology, which are ideally suited to the mass-production of filter devices required for use in television, offer elimination of the complicated and expensive filters previously used. Apart from reducing assembly costs and, more importantly, alignment or adjustment costs, mass-produced, integrated SAW filters give improved stability and performance and permit reduction in the physical size of the receiver.

Inexpensive microelectronic circuits allow the introduction of digital tuning for up to one hundred channels, so that one basic television receiver can be used in locations that receive signals from different transmitters. Such a tuning scheme, which is intrinsically electronic rather than mechanical, will simplify remote control. Indeed, research is under way to replace push-button and remote-control systems with direct human voice controllers such that the television would respond to a small set of spoken words. Stereo sound and stereo vision are other possible features. The technology for stereo sound transmission has existed for several years and stereo radio is commonplace, but stereo vision remains to be developed.

A demand exists for improvement in present television picture quality to create hi-fi television to be of comparable picture quality to that provided by the cinema, and indeed to that available from video disc players. Firstly, the possibility exists of improving picture quality within existing transmission standards – a system known as hi-fi 'A'. Here, by improvements in television camera technology as well as receiver improvements, the full potential of the 625-line system will be achieved using new digital filtering and suppression devices to eliminate many of the avoidable sources of picture degradation present in current systems. Use of charge-coupled device (CCD) integrated circuits offers a 25 per cent improvement in picture resolution, using CCD comb filters to isolate components of the video signal. A second system, hi-fi 'B', involves modification to the transmission standard such that specially designed receivers will be able to make use of the superior signal quality to produce an enhanced picture quality. The main constraint to full development of hi-fi 'B' systems is that any new transmission standards must remain compatible with existing standards to permit reception in unmodified receivers. Finally, there exists the possibility of significant improvements in picture standards using hi-fi 'C' systems, where the television signal is completely changed from its present form and television with 1125 lines and 30-inch tubes will be employed.

The television is increasingly being used as a viewing medium for use with non-broadcast information from video cassette recorders (VCR) or video disc players (VDP). The VCR can be used in the home for recording normal television programmes and can even record while the television is switched off or while another channel is being viewed. Inexpensive high-quality video colour cameras based on developments in CCD imaging circuits are beginning to emerge into the consumer market, and when linked to a light, portable VCR will herald the demise of the cine camera.

The video disc is pre-recorded in a similar fashion to an LP record and allows a wealth of information to be displayed at home for recreation, education and leisure. Several systems are competing to establish a standard but the Video Long Player system developed by Phillips appears favourite. In this system both the initial (master) recording and the reading during playback are performed optically using a fine laser beam. The separation between tracks on the disc is less than 2 μm and the focal depth is less than 1 μm, and extremely accurate and expensive guidance for the laser beam is required. Two playing modes can be used: fixed angular speed gives 30 minutes playing time per side but fixed linear speed allows an extension to 60 minutes per side, giving two hours of high quality television picture. The discs can be reproduced by pressing techniques similar to those used to produce long-playing records, and thus the production costs should be similar.

Microprocessor-based television game. Microprocessor-based television computer games connect directly into the aerial socket of any television and offer a new dimension in home entertainment. A range of cartridges is available, offering a broad selection of entertainment with action, adventure and excitement in combat games, sports games and mind games. The computer console plus controls costs in the region of $200, with games cartridges costing about $30. *Photograph courtesy of Ingersoll Electronics Ltd.*

One major use of domestic television that has received much publicity recently is the introduction of teletext and viewdata systems. Teletext is an information system comprising several hundred pages of text and graphics sent to the television receiver in standard transmissions on the normal television signal. Teletext pages are read from a computer memory and the page information is transmitted on unused (frame) lines in the television signal. At the receiver the desired page to be viewed is selected and held in memory, translated into characters and graphics form and displayed on the screen. Viewdata is a conceptually similar system but the signal is carried by cables and the user is not limited in the number of pages available from the large database

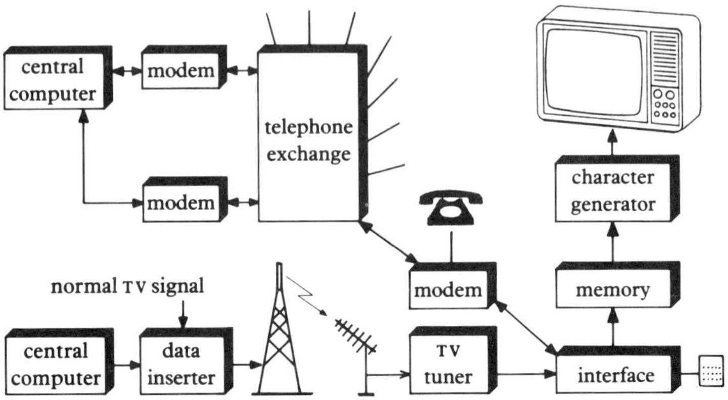

Teletext and viewdata systems. Teletext and viewdata are independent text and graphics display systems for home and office use. To achieve high market penetration the 'user' ends of both systems are identical, employing common LSI circuits. The means used to transmit data on the two systems are entirely different: teletext information is broadcast in the normal television signal while viewdata uses the telephone system. Viewdata systems allow interactive use of the data link and access to a virtually unlimited database.

held in the central computer of the viewdata system.

Britain holds the proud distinction of having provided the first public services in teletext and viewdata. Ceefax, the BBC's teletext transmission, and Oracle, its IBA equivalent, began broadcasting in 1976 after a two-year trial. Prestel, the British viewdata system, was opened in 1979. Teletext and viewdata are media that combine the news and features facilities of newspapers and magazines with those of television. Both are designed to employ common standards and to display text and graphics on standard television receivers. Like a newspaper, the textual and graphical information is organised into pages, with each page referring to the information presented on a single television display. The user can select the desired page by entering a page number into the electronics of the receiver. The pages can contain news, weather, sports information, advertisements, jokes, recipes or more specialised information such as financial reports, stock market positions, hotel reservation details or airline timetables.

Teletext is the generic term for a medium that carries broadcast text and graphics intended for display on a television receiver. The information is transmitted cyclically, one page at a time, so that the receiver must 'peel off' the selected page when it appears and store this page in memory for display. The data are digitally encoded and are usually sent out over the broadcast network as an accompaniment to the television signal.

The UK television signal is composed of 625 lines, divided when creating the picture into two interlaced fields of 312 and 313 lines respectively. A total of 575 lines are active in the picture, while the remaining occur in the field-blanking interval between successive fields. Two of these inactive lines in each field (four of the 625) are used to carry the teletext signal without causing interference to the normal television display. The teletext data is digitally encoded and superimposed on these lines for transmission at a bit rate of 6.9375 Mb/s. Each transmitted line carries the data for one complete 40-character row of the teletext display and each page is composed of 24 rows of 40 characters. Up to 800 pages can be included in a full transmission but, because the pages are transmitted cyclically, each additional page in the cycle adds to the average access time for any one page. Using two lines per field means that four rows are transmitted every 1/25 second, that is approximately four pages per second, and for the present system, which permits up to 800 pages to be used, the complete cycle time is 3 minutes and 20 seconds. Not all 800 available pages have to be transmitted on each cycle and, in addition, it is not necessary to transmit blank rows.

The 800 pages are partitioned into eight magazines of 100 pages each. This division is purely for organisational convenience; it does not affect the method of transmission and the primary advantage of this magazine arrangement is that it allows blocks of the national teletext transmission to be replaced locally on a region-by-region basis. Not all magazines need be transmitted, nor all available pages of each magazine.

The first row of each page is a page header containing the page number, display control codes and a time code. Subsequent rows are labelled with a row address and magazine address so that rows of different magazines can be interleaved as required in the transmission sequence. Pages that are likely to be frequently accessed can be transmitted more than once per cycle, reducing the average access time required by these pages by half or more (dependent on number of repetitions) at the expense of increasing total cycle time by only one quarter second. Pages that contain complementary information can be included under a single page number by transmitting one page for several cycles followed by its complementary page for the next few cycles. A good example of this is the Ceefax recipe page, where the ingredients and method of preparation are listed on separate pages that alternate when the recipe page number is selected. Finally each page can contain up to 3200 different sub-pages, addressable by selecting first the page number and then a time code, which effectively increases the addressing range. The time code sub-pages are also transmitted cyclically but now only one step of the time-code cycle is executed for every complete transition of the primary 800-page cycle. Such a configuration might be used to present, say, stock market prices, where a large number of pages are required to give a complete report. This extension permits a total capacity of over 2.5 million pages but the penalties in access time are obviously severe.

A number of different display modes allow varied presentation of text on each page. Every individual text and graphical character can be displayed in any one of seven colours: red, green, yellow, blue, magenta, cyan and white; each text character can be displayed on a background of any one of these colours and characters can be made to flash on and off. A special feature allows certain characters on a page to be concealed until a 'reveal' button on the user keyboard is depressed, two other keys cause the expansion of the bottom or top half of a page to double height for easier reading, and characters can be superimposed in boxed form on the normal television picture to provide newsflashes or subtitles.

All control, text and graphics characters are represented by seven bits plus an extra bit for detecting errors – a parity bit. The strategy employed by most decoders is to ignore characters that have errors, although more complex strategies, which integrate multiple receptions of the same page, have been implemented. Thus, on the first acquisition of a page, blanks will be left on the display wherever single-bit errors are detected. The subjective effect of such errors on the viewer is minimal. Because the probability of many errors on a single page is low, most omissions in text are easily interpolated. The most disturbing effect that may emerge from single parity error is a missed change-to-graphic or change-to-alphanumeric control character, producing strings of erroneous characters and upsetting an entire row. Nearly all errors are corrected the next time a page appears in the transmission cycle. Any new errors cause retention of the previous character so that most single errors are corrected in the second appearance of a page. More serious are the fortunately less frequent occurrences of double-bit errors in a single word, which not only cause erroneous text characters to be displayed, but can activate unwanted control modes even though the intended character may not have been a control character. These errors can be very difficult to interpolate and are likely to affect an entire row. The worst type of error that might occur would be complete loss of rows, interchanged rows or display of an incorrect page. To prevent this, all page header information and magazine and row address bytes are protected by codes that allow correction, at the decoder, of words with single-bit errors. Double-bit errors are rejected so that blank rows can appear on the first access of a page but erroneous rows will not be generated.

Microelectronics has had a phenomenal impact on reducing the cost and reliability of the hardware throughout the teletext system. It is widely acknowledged that implementation of a teletext system has only become economically justified and indeed feasible because of the advances brought by microelectronics technology. At the central computer, programmers format the pages on intelligent terminals with full keyboard and VDU facilities. Complete pages are passed to the back-up store of the main computer, where they can be inserted into the teletext programme when desired. Output pages are encoded onto the video waveform and passed down the distribution network for transmission across the airwaves or via cable links. At the receiving end a remote control keypad is used to select the desired page, which is then acquired, decoded and fed to the television display. The cost of the hardware that acquires, decodes and configures the teletext signal for television display represents the only expense to be borne by the consumer who desires teletext capabilities on his television receiver, and total decoder cost is therefore directly related to the number of probable users. It was clear from the outset of the teletext design programme that only a low-cost decoding system would generate the market demand necessary to make teletext feasible. Thus it is in the decoder design that microelectronics technology, recognised immediately as the only possible solution to the minimal unit cost problem, has made the greatest impact. Teletext decoder systems have been developed by several companies, including Mullard, General Instruments and Texas Instruments.

The Mullard infra-red controller employs two integrated circuits, one in the transmitter, the other in the receiver. Thirty-two commands are possible, comprising all standard teletext commands in addition to the channel selection and analogue adjustment controls found in ordinary television remote controllers. A single integrated circuit forms the basis of the transmitter section. The battery-powered handset contains a control keypad and an automatic power-up circuit, which keeps the controller on constant stand-by but at negligible current, until a key is depressed. Several external components are used to set the transmitted pulse width (typically 5 μs) and a driving transistor provides the forward biasing current to operate the light-emitting diodes (LEDs). When a key is depressed on the handset a short pseudo-random binary sequence is transmitted, followed by a seven-bit start code and a five-bit message. These twelve bits are then repeated and the receiver does not respond until all twenty-four have been received and checked. A silicon photodiode fitted with an infra-red filter receives the infra-red signal and another integrated circuit decodes the signals from the photodiode detector and performs error checks. The decoded commands are used to create a number of control functions. Four analogue outputs provide a direct output voltage, variable over 62 levels, to control such receiver functions as volume, brightness, colour and contrast. A separate output provides a mute facility to instantly cut out audio for teletext viewing. A 'clear' output and a 'step' output are used to drive the receiver's channel selector. When a new channel command is keyed a 'clear' pulse resets the receiver's channel selector to channel 1. A number of stepping signals are then sent to increment the selector to the commanded number. Two separate outputs are used to drive LEDs. One of these indicates that the receiver is still connected to the mains although it has been switched off by remote control; the other lights when a correct command has been received. The latter function is very useful, as page acquisition in teletext takes considerable time and can leave the user unsure if his command has been received by the teletext decoder. The teletext decoder features four custom integrated circuits: the timing chain (TIC); the video input processor (VIP); the teletext data acquisition and control (TAC); and the teletext character generator ROM (TROM).

The consumer interested in teletext is faced with two choices: adding an adapter to his existing television or purchasing (or hiring) a purpose-built receiver. The latter is certain to offer the better performance since receivers with built-in teletext decoders have been designed to provide optimal performance on both television and teletext display. This is not the case with most television receivers built prior to the introduction of teletext, since a number of receiver characteristics are more critical for teletext application than for normal viewing. Of paramount importance is the design of the intermediate frequency (IF) amplifier where the baseband video response and group delay should be essentially flat up to one-half the teletext bit rate. Automatic gain control (a.g.c.) systems that alter the IF bandpass shape must be avoided for teletext as they affect both amplitude and shape of the data pulses. Simple diode video demodulators used on older television receivers, and still found on some cheaper models, often introduce a great deal of non-linear distortion, which significantly reduces teletext decoding margins and signal symmetry. A poorly tuned local oscillator is also more critical to teletext performance than to ordinary video and thus the parameters of the automatic frequency control (a.f.c.) system are important. Furthermore, a teletext signal passing through a set-top adapter will, by definition, be subject to greater degradation on its path to the eventual display than will one which passes directly into a purpose-built receiver. The external adapter carries out all the same data stripping, decoding and video signal generation performed in the purpose-built set, but instead of directly driving the video amplifiers of the receiver the teletext signal from the adapter must then be encoded as a standard video waveform and passed through a further stage of video demodulation. This extra hardware also makes the external adapter relatively expensive. However, the addition of an adapter is always cheaper than purchase of a new television, so considerable attention must be devoted to minimising adapter cost if teletext is to be thoroughly successful.

In contrast to the teletext system, viewdata is a generic term for a medium of text and graphics transmission intended for colour television display in an interactive mode. To be interactive there must exist a two-way communication path between the viewdata user and the database for the system. In viewdata, when the user keys-in the required page number, his terminal sends this page number via a cable to the central computer, which then selects the information on that page and sends it to the user. Because of the interactive nature, viewdata cannot simply be broadcast, instead it is carried (usually) by the telephone network. The slow data rate of the telephone system is not a handicap since the pages are not transmitted cyclically, and unlike teletext there is no time wasted in waiting until the arrival of the slot in the transmission cycle on which the required page is placed.

The interactive capability of viewdata offers a number of advantages. Because specific pages are sent only on request, the number of pages available in the system database is restricted only by the size of the central computer – it is virtually unlimited. With teletext each additional page adds to the cycle transmission time, thus teletext tends to carry general interest information (similar to a newspaper) while viewdata is more likely to feature specialised information (similar to a magazine). The real power of the interactive capability of viewdata lies in the user's ability to communicate information other than page numbers to the computer. For example, the user can make a hotel reservation, request sales information through the post, order and pay for goods and play games requiring keyboard response.

Early on in the development programme for Prestel, the British viewdata system, it was recognised that there would be considerable demand for a receiver that handled both teletext and viewdata and that to be successful this receiver would have to be relatively inexpensive. It was therefore decided to employ maximum commonality between the two developing systems, with the maximum possible number of custom integrated circuits being shared between the two systems in the decoder and display hardware and with the displays of both using the same coding and character format. The sacrifices that this commonality of design inevitably entails have been mostly at the expense of viewdata. Because teletext is severely limited by access times, a much lower graphics and text display resolution was adopted than is technically possible with viewdata. Higher definition would have required substantially longer transmit times per page and would therefore have reduced the already small number of pages in the teletext transmission cycle. Viewdata is not intrinsically limited by any such constraint since pages are sent one at a time on request. Nevertheless, the sacrifice is, for the time being, small in comparison with the economic benefits of the common design.

In viewdata as in teletext, the user enters on a keypad the number of the page he wishes to see. However, since viewdata is transmitted over the telephone network, a connection must first be established betweeen the user terminal and the central computer. Like a telephone, each input/output port of the computer is assigned a number, which must be dialled or keyed by the user (many terminals do this automatically) and when connection has been established, a switch must be activated (on terminals where this is not automatic) to connect the user terminal modem to the telephone line, the purpose of the modem being to interface the digital logic of the terminal and computer with the analogue telephone line. When the user's modem is connected, a greeting frame transmitted by the computer appears on the television display and the user must then key in the number of the page he desires.

As with teletext, each transmitted character is composed of seven bits that define the character plus one even parity bit. The modem adds to each word a start bit and a stop bit and, although this slows the character transmission rate, it allows asynchronous transmission of the type commonly employed by existing modems in data communication links to be used with viewdata. Transmission to the computer takes place at 75 bits per second, a rate faster than the speed at which any person can manually key a command. The modem at the computer end transmits to the user at a rate of 1200 bits per second. Although these rates are slow by teletext standards, it must be remembered that access to a page in viewdata is virtually immediate on receipt of a page command so that the only waiting time is that time taken to transmit a page. With 960 ten-bit characters constituting a full page, the maximum transmission time per page is eight seconds. Individual lines can be written to the display as soon as they are received so that a picture is built up more rapidly than it can be read and annoying delays are not experienced by the viewer. The viewdata modems use frequency-shift keying (f.s.k.) to create audio frequency tones that define the logic condition of each bit. When no data are being transmitted the terminal continuously emits a 390 Hz carrier and similarly the computer emits a 1300 Hz carrier. These tones are detected in both modems to provide a continuity check. When data is being transmitted the carrier is frequency-shift keyed between the binary 1 and 0 frequencies as necessary.

These modulating frequencies are, for 1 and 0 respectively, 390 Hz and 450 Hz for terminal-to-computer transmission, and 1300 Hz and 2100 Hz for computer-to-terminal transmission. Duplex transmission allows simultaneous two-way transmission without interference. Characters keyed at the terminal are transmitted to the computer before they are displayed. On receipt they are echoed back to the terminal modem which sends them to the user's display only if they match the outgoing character. This provides some degree of error checking and assures the user that the system is functioning properly. Characters received for display also undergo Hamming or parity checks in the decoder. Errors in the viewdata are very rare due to the very slow transmission rates employed.

The display format, and therefore all row addressing, is the same as in teletext. The viewdata signal, of course, requires row addressing only and there is no need to extract signals according to, say, magazine addresses. All graphic and text character codes are identical to those used in teletext, but viewdata uses a separate set of control characters to cater for the different display and operational modes required by viewdata. The viewdata cursor appears as a bright rectangle occupying one complete character space on the display, to indicate the position of the next character to be written either by the computer or user. Appropriate keyed instructions or computer-generated control characters can be used to select or remove the cursor display or cause the cursor to flash on and off, and a counter keeps track of the cursor position at all times, whether or not it is displayed. Cursor-control characters can be used to manipulate the position of the cursor so that blank spaces need not be written to the display, allowing a page with a number of blank spaces to be transmitted with appreciably fewer than 960 characters, and therefore in less than eight seconds. The cursor is also useful as a visual aid to the viewdata user with a number of viewdata services requiring a keyed response beyond simple page numbers. The cursor indicates the position where the next keyed character will appear on the screen so that the user can keep track of, say, a message he is composing, and by manipulating the cursor the message can readily be edited.

An important feature of viewdata is the use of the existing switched telephone network. Since the cost of a connection through this network increases with distance, it is important that viewdata computers be sited as near as possible to potential users, thus a system like Prestel uses a distributed computer structure. Local computers are grouped under the control of regional centres where regional pages can be created, and these regional centres are in turn linked to the national computer centre where most of the information originates. Regional information can be sent to the national centre for distribution and each local computer stores all the national information as well as local information, so that connection to any computer allows the user access to a complete database.

Viewdata, with its virtually unlimited database, has tremendous potential as an office system to be used by businesses of all types and sizes. The different locations at which terminals are likely to be sited dictate the need for several different types of terminal. As with teletext, there is strong demand for external adapters to bring viewdata to existing televisions and for remote-controlled televisions with viewdata decoding circuitry built in. To meet business requirements a new range of dedicated viewdata terminals have been designed, and a typical business terminal is a desk-top unit intended only for close-range viewing – similar to the visual display unit of a computer terminal. Remote control is unnecessary, so the keyboard unit is part of the terminal itself. The interactive facilities of the viewdata service often request alphanumeric messages to be entered by the user, for example the user enters his address when booking a hotel. If this capability is to be most efficiently employed, the user terminal must include a full alphanumeric keypad. As this adds considerably to the terminal cost, the viewdata operation has been designed such that it can be entirely controlled by a standard numeric keypad only, with a full keyboard as an optional extra necessary only if message capability is desired. Most current viewdata terminals require only a single key to be depressed to establish connection between the terminal and computer, removing the need for the user to dial manually the number of the local computer on a telephone set and then switch operation to the modem.

A number of manufacturers have developed combined teletext and viewdata decoders. The Mullard system previously described for teletext requires the addition of a viewdata acquisition circuit (VAC) and a line coupling unit (LCU) for operation with viewdata. The main function of the VAC is to write viewdata into memory under the control of an MP8084 microprocessor operating with one kilobyte of internal memory and two kilobytes external ROM. A microprocessor has been employed so that, by simply replacing the ROM, future viewdata developments can be catered for by the existing decoder.

The 24-hour Prestel viewdata service is now available to (geographic) sections of the public, but growth has been much slower than that of teletext, primarily because expansion of the system is capital-intensive. To open Prestel to a new group of users entails the installation of an entire telephone-compatible computer system. By 1981, 60 per cent of UK telephone users were within local call fee access of a Prestel computer, but concentrating development on the larger population centres and industrial areas will mean that the percentage of major businesses covered is much greater. The Prestel database currently has capacity for 250,000 pages, of which over 200,000 have already been contracted, and over 150,000 of these are currently accessible.

Prestel offers such a wide range of information that the only way to appreciate its scope and diversity is by reference to its directory, where subjects covered range from abortion to zinc. Railways and airlines list complete timetables and regularly updated accounts of seat availability; retailers offer price lists; businesses present their annual financial reports; societies provide biographical details and agenda; a range of statistics ranging from world records to metric equivalent charts is offered; detailed news on specialist subjects abounds. The most interesting pages of Prestel are those that fully use its interactive capabilities. Termed 'response pages', these are the pages that require input from the user. Armed with a credit card one can order a variety of goods by keying in a number assigned to the desired item, followed by the credit card number. A myriad of free literature can be requested merely by entering one's name and address, when the Prestel computer stores the request so that it can be passed on to the relevant information providers. Sequential games such as 'Mastermind' and 'Scissors, Paper, Stone' can also be played.

Pages are leased at a fee and it is up to the lessee to recoup this fee by charging the user for page access. The owner of a terminal can set his own password to prevent the terminal being used outwith his authority and this password can be changed at any time, following the directions listed on a special password page. Many information providers may want their information to be available only to a select group of users, a feature available on Prestel where certain pages can be made accessible only to 'closed user groups' by requiring that a special identifier be incorporated in admissible terminals. Alternatively, or additionally, a password can be employed so that accessibility can be restricted by terminal, or user, or both.

Significantly, all major British television manufacturers are now producing receivers for teletext and viewdata, and the large microelectronics companies are involved in continuing programmes to develop new and improved silicon integrated circuits for these systems. We have now reached a position in the evolution of the television receiver where the domestic television can yield a whole new range of services and facilities where, by connection to teletext and viewdata systems, the television will form an integrated home terminal system. Such a system could monitor many functions in the home, such as domestic appliances and heating and ventilation systems. Energy management can be simply implemented using this integrated home terminal, as can education, information processing, entertainment, communications, control of the domestic environment, security, accounts and recipe listing. These developments dictate that the domestic television receiver will remain the dominant consumer product based on microelectronics, certainly into the foreseeable future.

Microelectronics has created an enormous consumer market for intelligent and educational games. Some products that have emerged include 'Little Professor' and 'Data Man', elaborate calculators that teach mathematics to young children, and 'Speak and Spell', which teaches the young child to spell words in response to a synthesised speech output, asking the child to spell each word in turn and allowing two attempts at keying-in an answer before providing the correct answer on the display. Talking products are recognised as the next major growth area in the consumer market and several speech synthesis circuits have been developed.

Speech generation using digitally stored speech waveforms – a direct replacement for normal tape recorders – would require enormous amounts of memory. For example, to store three minutes of speech using pulse code modulation (PCM) would require approximately 13 million bits of digital storage. Speech synthesis techniques offer a convenient alternative to the high memory requirements of such waveform encoded systems. These techniques use a mathematical model of speech production with the voiced sounds of speech, such as 'a' and 'e', being synthesised by connecting a periodic source generator to a model of the vocal tract. Unvoiced sounds, like 's', 'f' and 'sh' use a random source generator. The difference between the various approaches lies in the vocal tract models used.

Microelectronics in learning aids. Developed as a learning aid mainly for children of seven years and over, the 'Speak & Spell' product actually asks verbally and then provides solutions to spelling questions. Its main value in teaching lies in the fact that the child's interest and confidence in spelling and pronunciation can be developed at the child's own pace. The basic computer memory contains 240 problem words arranged in four levels of difficulty, and plug-in word modules are available to expand the power of the product to specific areas in spelling such as words that sound alike but are different in spelling and meaning. *Photograph courtesy of Texas Instruments Ltd.*

The technique known as 'linear predictive coding' synthesises the speech signal using a time-varying filter to represent the vocal tract. This technique is used in the 'Speak and Spell' product, which has faced certain stringent constraints in its development. Firstly it was required to be produced cheaply to attract mass market sales and secondly, since it was to be used to teach children spelling, language, and phonetics, the quality of synthesised speech had to be high. In the method of linear prediction each speech sample is predicted as a result of previous samples and the digital filter output samples are converted to analogue form using a digital-to-analogue converter. One of the most attractive features of linear predictive coding is the natural quality of the speech so produced. The processing required to produce synthetic speech is complex, even for the restricted basic 200-word vocabulary of the 'Speak and Spell' product. Four individual integrated circuits are employed: the speech synthesiser, which contains the source generators and the digital filter; two (ROM) memories, which hold the required filter parameters for given speech sounds; and the controller chip. The specific vocabulary can be changed using interchangeable solid state speech memory modules. The 'Speak and Spell' digital vocabulary is generated from high-quality tape recordings, with each word being digitised at an 8 kHz sample rate for analysis in a minicomputer to determine pitch and vocal tract parameters for use in the digital filter of the synthesiser chip. The next step is to code each parameter for storage in memory. The resulting data rate is 1133 bits/second.

A second speech technique, known as formant synthesis, reconstructs the time-varying frequency parameters of speech – the resonant frequencies and energy levels. Low storage requirements can be achieved since the spectral parameters of speech change relatively slowly. Formant vocoder chips also contain two signal sources, a variable oscillator to simulate the vocal chords and a noise generator. The outputs of these sources are shaped by a bank of parallel filters, which simulate the cavities of the vocal tract. The ultimate in low bit-rate speech synthesis is the phoneme synthesis method, in which the synthesis of speech is controlled by parameters derived by separate sounds – or phonemes – present in speech.

In addition to their use as learning aids, speech synthesis chips are used in many consumer products. For example, a talking microwave oven, which can provide spoken cooking information for a specific food, is available. Other applications exist in speaking chess or bridge games and other entertainment products, as well as the use of speaking products in the motor car to provide the driver with an audible warning of hazards, faults or problems. The development of this market will create an ancillary requirement for systems that can prepare the speech data for a given product and, as the demands for speech synthesisers increase, improvements in data preparation will take place and speech analysis and word recognition systems will eventually become as commonplace as speech synthesisers. These developments will ultimately create a further new area of the consumer market, with telephone sets that will dial a number in response to the number being spoken as well as new products for control of the household duties under spoken control.

Microelectronics has been introduced into domestic aids in the kitchen, and white goods products such as cookers, washing machines, dishwashers and sewing machines that incorporate microelectronic circuits are now available to ease the load on the busy housewife. With the advent of a multitude of new fabrics, finishes and treatments, and a wide selection of washing requirements and available detergents, clothes washing is no longer a simple matter. Fortunately, there now exists an international standard coding system for washing requirements and based on these standards a set of basic wash programmes has been evolved, each defining the various wash conditions such as temperature, wash time, agitation and water levels for the various stages in the wash cycle.

A range of integrated circuits is available for control of domestic equipment, based essentially on the microprocessor to provide the complete programming and timing control required by modern products. The different functions of the machine being controlled are held in memory (ROM), enabling the one basic controller circuit to control a variety of machines including driers,

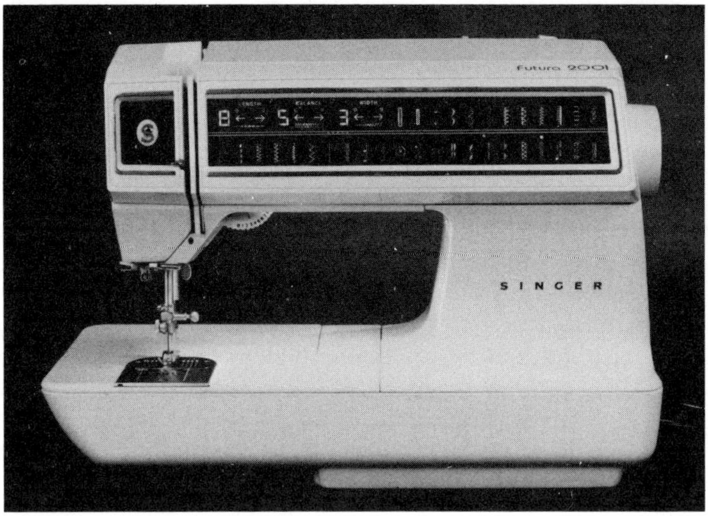

Electronic sewing machine. By incorporating a microprocessor to control the timing and sequencing of operations required in a normal sewing machine it becomes simple to select any of 27 different functional and decorative stitches plus two buttonhole sequences, simply by touching the relevant symbol on the illuminated digital control panel of the sewing machine. Similar touch control can be used to select a new stitch length, width and balance or to produce a mirror image of a stitch pattern. In addition, a warning light glows when the bobbin thread needs renewing. Use of the microprocessor replaces a total of 350 electromechanical components, making the product more compact, more reliable, and simpler to maintain. *Photograph courtesy of The Singer Company (U.K.) Ltd.*

heaters, air conditioners and cookers, as well as dishwashers and washing machines. For washing-machine applications the controller circuits make use of the trend towards standardised wash cycles, while at the same time incorporating maximum flexibility for manual override by the user when necessary and for automatic dispensing of various wash detergents, bleaches and softeners as desired by the user. Further, the microprocessor-based controller can also incorporate safety interlocks, such as door switches to prevent loading doors being opened during spin cycles or when the water level in the machine is high. In washing-machine controllers the memory contents are divided into nine sections, each controlling a separate wash programme. The data from the memory is selected under the control of a sequence counter, which steps through 20 positions to define the programmed steps in the wash cycle. The sequence counter moves to its next position on completion of the activity specified by the current step in the wash cycle, for example, when the desired water level has been reached, when the water temperature has been reached or when the time allocated to the step has elapsed.

It has been predicted that the failure rate of microprocessor controlled washing machines will be essentially similar to the failure rates of existing electromechanical controllers. In addition, the costs of the microelectronic controller will be comparable with those of the existing controllers. However, the user-oriented features of apparent 'intelligence' in the washing machine, coupled with the more futuristic control and display potential of the machine front, will undoubtedly lead to wide acceptance of such microprocessor controlled appliances.

Microelectronic applications in the motor car can be divided into two main categories. The first is engine management, where the aim is to variously improve engine performance, reduce fuel consumption, reduce exhaust levels and prolong engine life. The second category is the various dashboard display and information features, plus several new warning and safety systems that are essential selling points for the product.

Microprocessor control of the car engine, electronic engine control (EEC), is currently at an advanced level in the United States, where there exists a legal requirement to conform to strict government limits on exhaust gas pollution coupled with a required minimum fuel-economy level. The low pollution and high fuel-economy levels set by the US Government cannot be met simply by mechanical control systems, and manufacturers have turned to microelectronics to meet the required standards. It is worth noting that exhaust emission regulations in Europe are considerably less severe than in the United States.

The most sophisticated microprocessor-based engine management system currently available is the second generation of Ford Electronic Engine Control, EEC-2. This system incorporates a twelve-bit microprocessor and succeeds the EEC-I design, which controlled only ignition advance/retard levels and recirculation of exhaust gas back into the engine. In addition to these two features, the EEC-2 system controls the air-to-fuel ratio (mixture) by incorporating an electronically controlled feedback carburettor. Inputs to the EEC-2 system include manifold pressure, engine coolant temperature, crankshaft position and exhaust gas content. Three basic look-up tables (ROM) are stored in the system to model the engine system of the car such that engine performance can be gauged in comparison to the tables. The specific engine information contained in the memory relates to air-to-fuel ratio, spark advance and exhaust recirculation rate. The microprocessor, presented with such fundamental parameters as manifold pressure and crankshaft speed, selects the optimum control point for the spark, air-fuel mixture and exhaust gas recirculation from these tables, and the appropriate microprocessor decisions are used to control a set of modulators and amplifiers, which change the engine conditions via a set of actuators. For example, the feedback carburettor is controlled by a stepping motor that adjusts the fuel-metering rods of the carburettor.

The spark advance, fuel control and exhaust gas recirculation controls are all contained in a single integrated circuit. The spark advance function requires information on piston position and crankshaft speed and given this input data, the controller can provide the correct ignition advance setting for these conditions from the tables stored in memory. It is possible to extend this control to the condition where the car is cruising at constant speed such that, for fuel economy, it is possible to arrange that the engine does not fire on all cylinders. A calibration module within the EEC-2 system permits the basic controller to be used with various models of cars with different characteristics such as different power-to-weight ratios, engine size and gear ratios.

In addition to engine control, microprocessors can be used for other tasks in the car such as automatic radio channel selection and more informative and more easily readable instrument panels. An example of this is a trip computer, which can provide the driver with a display of average speed, fuel consumption, estimated arrival time, and time of day. As the use of microprocessors becomes more common in motor car design, the trend will be towards use of several microprocessors distributed throughout the car, one for engine control, another for the electronic dashboard and display control, and other for radio tuning, transmission control and engine diagnostics. There are several reasons for this. First, separate systems lead to improved reliability, because if one processor fails, it does not affect the others. Secondly, as the cost of microelectronics decreases, it becomes less expensive to distribute the microprocessors around the car than it would be to run extra cables through the wiring loom to carry information to a centralised microprocessor. Thirdly, the most important in the short term, many of the available microprocessor systems are options that can be sold separately.

One obvious use of microelectronics in the motor car will be towards provision of improved safety. Examples here include anti-skid computers, which will adjust braking pressure to maintain the optimum anti-skid conditions between the road surface and each individual wheel, and brake-light flash circuits which will 'pump' the brake lights at a rate proportional to the rate of deceleration of the car, producing an effective visual warning to following drivers.

6. MEDICAL APPLICATIONS OF MICROELECTRONICS

The rapidly falling cost, coupled with the increasing sophistication, of microelectronics makes this technology ideally suited to new developments in medical equipment ranging from the small desk-top computer system in the doctor's surgery used to store patients' case histories, to the sophisticated computerised body-scanning equipment of the specialised hospital department. The range of applications of microelectronics in medicine is vast, but the field can be divided into two main areas – examination and monitoring of the patient, and treatment of the patient.

Attempting to discover what is going on inside a patient presents physical difficulties for which there is often no simple engineering solution. There is no circuit diagram or technical manual for the individual patient, although general information has, of course, been assimilated by the doctor as a result of his education and experience. The doctor must therefore rely on information given to him by the patient and on a number of tests, which he can carry out as appropriate.

The use of computers for direct interaction with the patient is an expanding area of interest, and a number of developments have been reported. Initial fears that patients would not trust the computer used as an aid to diagnosis, and would be unwilling to provide accurate responses to very personal questions, have proved groundless and, indeed, with computer programs that are patient-friendly, patients appear willing to give personal details to the computer without any of the embarrassment that they might feel when speaking to another human being. The standard of diagnosis given by such computer systems is high, and compares well to the diagnosis of experienced medical practitioners used as a check.

The options normally open to an engineer of opening up and probing inside a piece of equipment that requires servicing, or even of switching off the equipment to permit removal and test of some individual part, are not applicable to the medical practitioner. Indeed, only when it is too late can a complete internal physical examination take place. What methods, then, are available to the doctor to perform a physical examination of the internal workings of the human body and how can microelectronics enhance these methods?

The insertion of narrow flexible probes is one commonly used technique to gather information without excessive disturbance to the patient. The probes may only require to enter the body a short distance, for example when used to intercept a vein or artery in order to measure blood pressure. The pressure of the blood stream can then be relayed to a small electronic sensor mounted outside the body, thereby providing a dynamic electrical signal to give an indication of the blood pressure. Similarly, methods exist for measuring the amount of carbon dioxide in a patient's breath, based on the fact that when carbon dioxide gas is irradiated with infra-red radiation it absorbs the energy and expands in volume. By an appropriate sequence of valve operations, this expansion can be sensed by an electronic transducer, and the quantities being measured can be represented by electrical quantities for processing by microelectronic circuits. We are now in a position to see how microelectronics can be usefully applied, in these applications where the signals must be amplified, transmitted, processed and displayed before the doctor is able to take any decision affecting the patient. These processes are, of course, already carried out using some sort of electronic systems. The introduction of microelectronics will make the systems more reliable, with more features, and possibly cheaper.

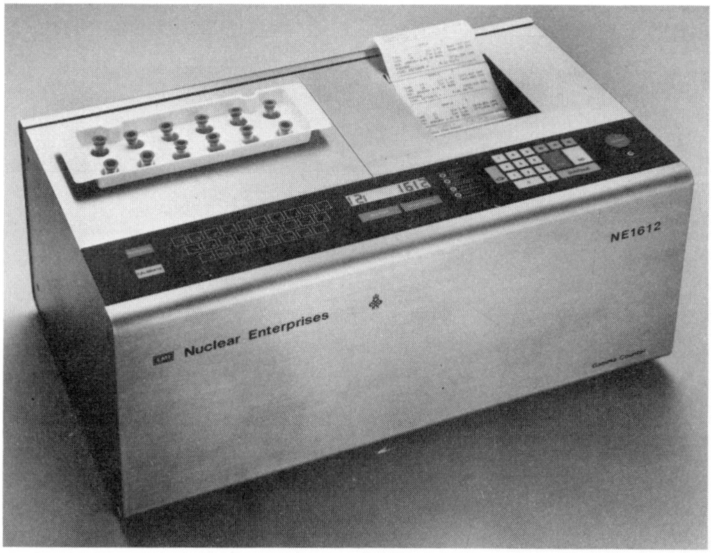

Microprocessor-based gamma counter. The gamma counter provides a sensitive and accurate means of measuring the radioactive content of medical samples. Examples include the screening of blood samples for transfusion and in dosage definition for drug administration. Incorporation of a microprocessor allows the user to define the required parameters interactively via a special keyboard, and the microprocessor can be used to compute the count level and display the information numerically or graphically by means of an integral printer. *Photograph courtesy of Nuclear Enterprises Ltd.*

A wide variety of transducers are available to measure such parameters as temperature, pressure and vibration and special-purpose integrated circuits have been developed to enhance the performance of these transducers. As these integrated circuits are very small, they can be attached to or even inserted into the patient. The fact that the electrical signal receives its first processing step so much closer to the origin of the signal significantly improves the quality of the result. A common feature of the signals available at the transducer – for instance when detecting the surface vibrations associated with the beating of a fetal heart – is that it is confused, subject to several sources of interference (e.g. the mother's heart), and rather weak. Therefore, the signal processing that would previously have been carried out in the doctor's brain when listening to a stethoscope, can now be done by electronic means. One very direct way of gaining information on the inside of the body, without any signal processing, is the fibre-optic probe used to transmit an image directly to the doctor. These probes are only a few millimetres in diameter and so can be inserted into many previously inaccessible body cavities. However, they cannot themselves be described as microelectronic devices, although they may be used with microelectronic systems for displaying and recording the images produced.

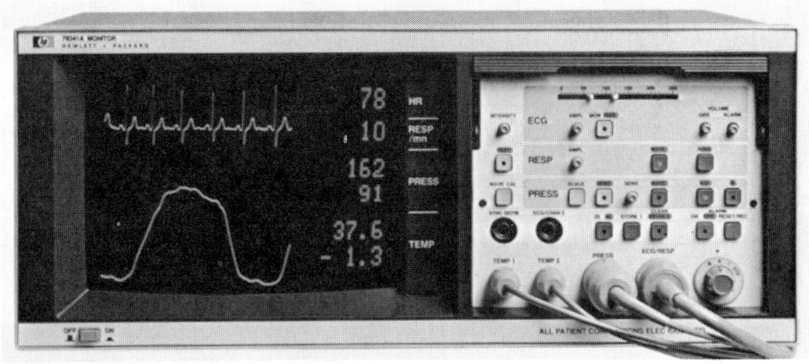

Patient monitor systems. Compact, microelectronics-based patient bedside monitors provide a clear, immediate indication of vital patient data. The system shown here (HP 78341A) monitors such data as ECG (top trace) and blood pressure (bottom trace) and can also compute and display numerical values of heart rate (here 78 beats per minute), respiration rate (10 per minute), temperature (37.6°C) and blood pressure (162/91). Limits for these monitored parameters can be programmed such that when the patient's condition reaches a critical state, a warning will be automatically sounded. *Photograph courtesy of Hewlett Packard Ltd.*

The body itself produces some electrical signals without the need for transducers. Nerves produce small electrical signals, but these can only be efficiently detected with fine needle probes carefully positioned in the nerves and special microelectronic amplifiers will no doubt make this a more routine procedure. In addition, the muscles activated by these nerves also produce an electrical signal that can be detected by electrodes placed on the surface of the skin, thereby requiring no implanted electrodes to produce a signal known as the electromyographic (EMG) signal, which is related to muscle activity. This is to be distinguished from the electroencephalogram (EEG) derived from electrodes placed on the head to detect nerve activity in the brain. The EMG signal has two important applications, one of which is the activation of artificial limbs and which will be described later. The other is the very important diagnostic tool of the electrocardiogram (ECG).

When used simply for patient monitoring, the ECG can be derived from just two electrodes attached to the patient's chest. Problems of interfering signals, e.g. from extraneous sources such as the mains supply, or internally generated signals like the patient's breathing, together with the consideration of the patient's safety from electric shock, put very stringent requirements on the performance of the ECG amplifiers. Here is another application for microelectronics to replace existing amplifiers and also to enhance system performance by including such features as analogue-to-digital converters on the same chips. The placing of electrodes on the chest also allows the measurement of respiration, by passing a very small current through the body and measuring the time-changing impedance. With sufficiently sophisticated circuitry and signal-processing, the same electrodes can be used for monitoring both respiration and heart activity.

The general form of the resulting ECG signal shows distinct peaks. When the regular rhythm of the heartbeat is disturbed it is most important to be able to distinguish whether the form of this disturbance, or arrhythmia, is of major or minor significance, so that any necessary action can be taken as soon as possible. A number of microelectronic monitoring instruments have been developed that can detect dangerous irregularities in the heartbeat and give appropriate automatic warning to supervising staff, as it is obviously not possible to have staff continuously monitoring these waveforms by direct visual observation. Typically, 99 per cent of potentially serious arrhythmias are detected by such systems, leading to immediate treatment in 95 per cent of the cases compared with a figure of only 17 per cent using conventional methods. Individual bedside monitors cost in the region of $5000–$10,000 and have hitherto used mostly analogue signal-processing circuits, although this is now changing rapidly with the availability of cheap computer power in the shape of microprocessors. These monitors learn the normal ECG of the particular patient and subsequently monitor the incoming ECG for certain deviations. The electronics should ignore slow changes, interference due to patient movement or mains interference and changes in the amplitude of the signal.

An alternative approach is the use of a 'multibed' system, in which the ECG signals from a number of patients are fed into a central monitoring system. Concentrating the signal processing into one unit permits use of a more sophisticated processor, which can extract more information about the condition of the patients. The processor can incorporate a small digital computer, which will compute fairly complex algorithms for the detection of arrhythmias and the prediction of trouble. They can also record the progress of patients for subsequent investigation by cardiologists.

In order to gain information on the disposition of the various parts of the body, a great deal of effort has been expended over many years, culminating in a range of very expensive equipment that offers the attraction of being non-harmful to the patient. The original means for non-invasive internal inspection, still very much in current use, is the X-ray. This involves the use of electromagnetic radiation, like radio waves and light, but of extremely short wavelength and hence able to penetrate matter. Everyone is familiar with the usual X-ray shadowgraph of the lungs, head etc. However, such shadowgraphs are fairly crude two-dimensional representations of a three-dimensional object

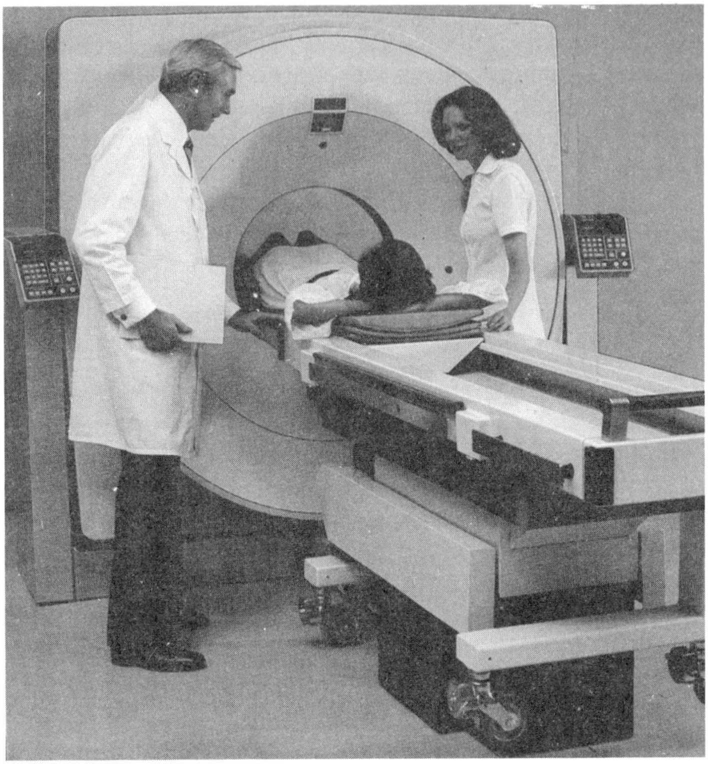

Computerised X-ray scanner. By making the X-ray beam scan over the patient, the computed tomographic (CT) system allows non-invasive visualisation of internal body elements including the brain. The display takes the form of a two-dimensional slice through the body and the high quality images with excellent spatial and contrast resolution are created from some 1½ million readings taken during each three-second body scan. Advances in X-ray sensor technology have led to greatly increased photon detection efficiency giving diagnostically excellent images at lower radiation levels. *Photograph courtesy of International General Electric Ltd.*

and give only very slight information on different sorts of tissues. No real advance was made in this field until the introduction of the X-ray tomograph in 1972, and this development only became possible because of the enhanced computing power available as a result of the introduction of small- and medium-scale integrated circuits. It is regrettable that the price of these machines continues to escalate while the cost of electronics continually decreases. This is because of the increasing sophistication of the non-computing parts of the machine in order to satisfy user demands for more precise details in terms of contrast and spatial resolution, together with faster production of results.

In order to appreciate the complexity of the problem, we need a rough outline of how these instruments achieve their objectives. It is not possible to focus X-rays in the same way as light, so the best that can be done is to place apertures in front of the X-ray source to collimate the rays into a reasonably narrow beam. When the X-rays have passed through the patient, they are detected by a scintillator/photomultiplier combination, where the X-ray energy is first converted to pulses of light and then to electric current. In the single transmitter/receiver pair rotating round the patient, as was the case in early instruments, body regions of greater density attenuate the signal more than those of low density. By gathering data at all possible angles of rotation, it seems intuitively reasonable that a unique description of everything in the scanned slice could be produced. The problem was how to interpret the information to produce an accurate image showing the correct densities at all the correct points. This is essentially a problem of solving thousands of simultaneous equations and so is ideally suited to a digital computer.

X-ray systems now have multiple detectors to speed up the process, but at a considerable cost in complexity, as all the detectors used must be constantly checked and adjusted to have identical sensitivity. About 30,000 readings, each represented by a sixteen-bit number, are taken on scanning each slice of the patient, and this can be completed in about two seconds in current machines. The computer can be used to display the results in a number of different ways. The cross-sections themselves can be displayed, using different colours to indicate different densities and with facilities to compress or expand contrast as required. Alternatively, the information from a number of different slices can be combined such that a three-dimensional impression of the inside of the patient can be produced. The spatial resolution of the technique is about 0.75 mm. The sixteen-bit data from the tomograph, being in digital form, can be transferred directly to computers, which can also be used in calculating the dosage and positioning of X-rays used in the treatment of certain conditions, thereby speeding up the process and adding further precision.

X-rays are harmful in large doses, so that some care must be used n the application of any X-ray technique. However, as the detectors used are 95 per cent efficient, there is less damage caused by a tomograph machine than by a conventional X-ray using film as the recording medium. Another method that uses potentially harmful radiation is the gamma camera, which relies on the introduction of mildly radioactive material into the patient. The radioactive isotopes used have short half-lives so that potential hazards are minimised. During the time when the consumed material is active, the radiation is detected using a sophisticated imaging system with multiple detectors coupled to a special digital computer.

Ultrasonic scanners have the advantage of not damaging the patient in any way. The ultrasonic energy can be focused and is usually employed in a radar-like pulse-echo mode. Mechanical scanning, achieved by the operator moving the transmitter/detector head, is often employed. This method is not, however, capable of producing results having the same quality as the X-ray computer tomograph.

The most recently developed technique is that employing nuclear magnetic resonance, which is completely non-harmful and gives promise of producing

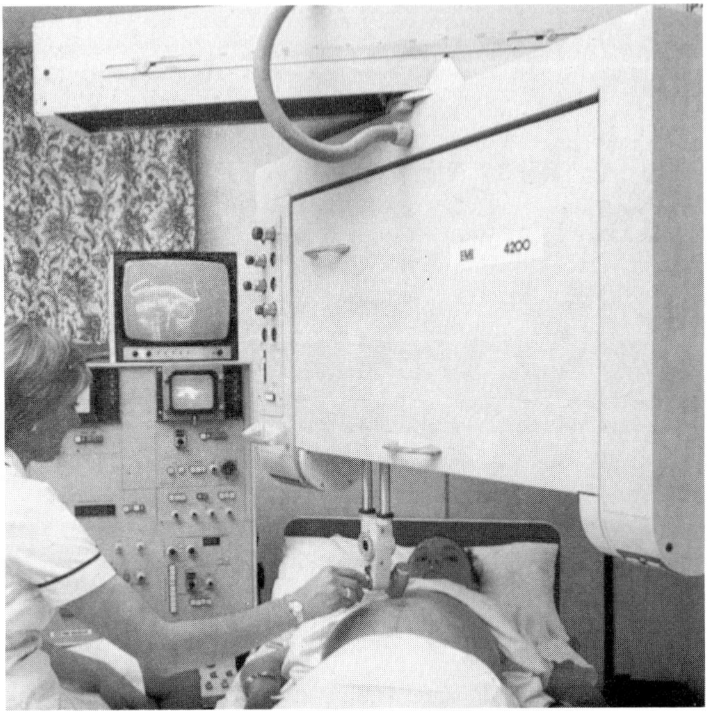

Ultrasonic scanner. Use of ultrasonics as an alternative to X-ray radiation offers a system which is intrinsically safer and the frequency of operation can be selected in the region 1.5 MHz to 10 MHz to allow the operator to optimise the system performance for a particular examination. The scanner operates by measuring echoes from organs in the body; a large echo is displayed as a white region, a small echo as a dark area, and the time of arrival of the echo provides positional information. *Photograph courtesy of Fischer Ultrasound Ltd.*

very high-quality results. It detects the protons in the nuclei of atoms by detecting the particular frequency at which they resonate when placed in a magnetic field. Protons have a spin and a magnetic moment, so that when placed in a magnetic field the axis of spin precesses or wobbles, and energy is therefore emitted or absorbed by the atom at this (radio) frequency. Hence it is necessary to immerse the patient in a static magnetic field with a superimposed radio-frequency exciting field. It is then possible to vary the steady field so that a small region of the correct field strength is scanned through the patient. As this is done, the r.f. field is detected and gives an indication of the proton density at each point. At the present time, recording times of up to two hours are necessary to give the best imaging. Producing the large magnetic fields needed to accommodate the human body poses some considerable engineering problems, and superconducting magnets (cooled with liquid helium) are being used.

The computing and control power of microelectronics has made these scanners possible and given them many useful features, but microelectronics cannot make these systems cheap and readily available because of all the other expensive engineering involved, and because of the skilled personnel needed to operate them and interpret the results. Scanners can be of enormous value both in revealing conditions at an early stage and in revealing conditions that could not otherwise be detected, but they cannot in themselves cure anything, and may often simply reveal a condition that is currently incurable.

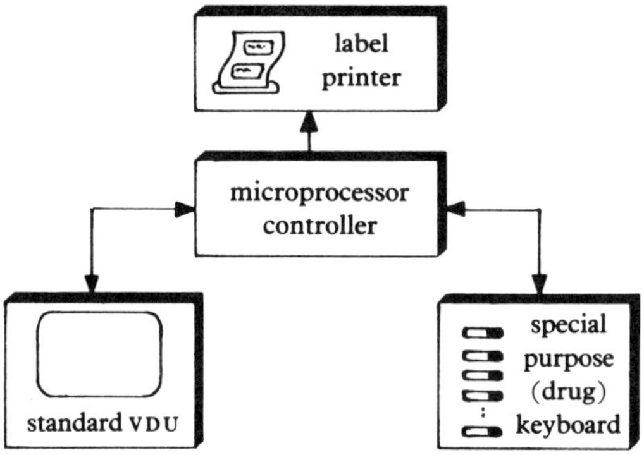

Prescription label processor. Some 92% of all prescription items can be covered by a total of 40 drugs, thus it is feasible to produce a special purpose keyboard where buttons correspond to drugs, dosage and other special instructions. The microprocessor acts as a controller for the keyboard, printer and VDU. The processor operation is simple and requires no operator training.

One area of medicine affected widely by microelectronics is the use of computers for the administration of treatment. A good example of this is automatic label processing, which removes a great deal of routine work from the hospital pharmacy by automatically labelling, under microprocessor control, the required prescription items, with each container correctly labelled for the individual patient. A special-purpose machine based on a Motorola 6800 microprocessor is available. Use of a microprocessor-based system was chosen in preference to the many standard microcomputers available mainly because of the requirement that the existing staff should be able to use the system with the absolute minimum of instruction and with no knowledge of computers or computing techniques. To facilitate use of the system with existing staff a special purpose keyboard was introduced, which has a separate key for each of the 40 most commonly prescribed drugs, covering 92 per cent of all prescriptions. The remaining 8 per cent are only required on a few occasions each month and are covered by a standard keyboard. Most of the information, which previously had to be individually typed, is therefore now held in computer memory and is printed out automatically with much less labour, with the result that the time spent in labelling has been reduced by about two-thirds.

As with all organisations, it is obvious that computers can make a considerable contribution to hospital administration, but it is also possible that they may also become of use to the GP. With appropriately tailored software the doctor's records of patients could be computerised, and so could be updated more easily and become more rapidly accessible. When all this information has been stored in computer memory, it is conceivable that it would be worth processing it to see if patterns could be deduced about the incidence of various diseases and hence eventually to the possibility of prevention or cure. However, any such system would need a policy decision on the part of the National Health Service to ensure uniformity and the collation of data on a national scale. The equipment must be very carefully designed so that it requires no computer expertise in the user, and it has to be clear to the GP that he would gain from using the system rather than having to spend a painful period of adjustment to it.

Another area of patient treatment in which microelectronics has a very clear part to play is in the provision of devices to aid the disabled. Artificial limbs are one major example, where microelectronics leads to significant improvements in the field of powered prostheses. One of the major problems with artificial limbs is that each patient needs an appliance directly tailored to his particular disability. There are no universal solutions.

A great deal of work has been done in the last 25 years concerned with the provision of artificial arms and hands, primary attention being focused on the provision of hands that function reasonably well and, very importantly, have a reasonably normal appearance. If this condition is not met there is a danger that the patient will not wear it, even though it functions quite well. The actuating signal is derived from the electromyographic voltage produced by the muscles in the remaining part of the patient's arm and since this signal is small the first step is to amplify it. As it is subject to various sources of noise, as discussed earlier, it must then be processed to eliminate the noise signals. Both of these processes are ideally suited to microelectronic devices, which are small and use little power and can therefore be sited inconspicuously on the limb. The signal is then applied to control the power source, and there are significant problems in storing power in a form that the patient can easily carry. In order to save power, it is possible to use devices such as elbow locks but in any event it is necessary to have either a re-chargeable battery pack or a carbon dioxide cylinder to provide the considerable energy required to move a limb or grip an object with a hand.

Lower-limb prostheses are very complex mechanically, and in order to produce a series of movements that simulate normal walking fairly sophisticated control is needed. Use of a microelectronic memory programmed from life appears to be indicated here. Each patient having a microelectronic-controlled prosthesis can have his individual memory chip programmed to suit him, thereby possibly introducing somewhat greater standardisation into the rest of the system.

Devices are also available to people with varying degrees of paralysis to enable them to carry out certain tasks. The wheelchair patient nowadays has mobility at the touch of a lever on his electric wheelchair. Modern everyday objects can be used directly or with little modification, such as remote control television and video recorders, electric typewriters; all employing microelectronic devices. For the deaf some success has been reported towards development of a device that will instantly convert the spoken word into a written form. Such a device will be extremely complex, and will only be available in a form that is economically feasible and of acceptable size by the application of microelectronics technology.

For the blind there is less in prospect. There are ultrasonic and laser devices to aid mobility but no hope exists for restoration of reasonable vision for a long time to come. Aids to reading ordinary printed text are available, so that the blind person is not restricted to reading books printed in Braille. One such device fits on the reader's finger and converts each letter on the page to a tactile form. The shape of the letter is sensed by an array of 144 phototransistors and the signal from each of these is transmitted to a piezoelectrically stimulated rod in contact with the finger. It takes some time to learn how to use such a device effectively, and further development demands incorporation of sufficient computing power to give direct speech output. Very little is really feasible in the way of a bionic man, where a lost sense is replaced by some sort of implanted device. However, in some cases of loss of hearing there have been some successes with cochlea implants, in cases where the hair cells have been lost but the nerve fibres are still intact. There are defects in the sounds that the patient hears, but it is likely that microelectronics signal processing could make significant improvements. Direct implants to the brain have not yet been used for this purpose.

Experiments have been done on implanting electrodes in the patient's visual cortex in order to give some restoration of vision. The patient sees flashes of light when the electrodes are stimulated and can see patterns when a suitable pattern of the (64) electrodes are stimulated using a large computer. It is expected that the technology will eventually be developed to the point where thousands of electrodes will be implanted and stimulated by a microcomputer that picks up a video signal from a microelectronic camera small enough to be fitted into a patient's eye socket.

7. MICROPROCESSORS IN INDUSTRIAL PROCESS CONTROL

Successful production, which can be defined as 'the speedy and efficient manufacture of goods of the required quality and cost' is not simply a series of individual operations performed to convert raw materials to a final product. Many functions, such as machine monitoring and stock control, which are commonly thought of as being peripheral to the actual manufacturing process, are equally important for achieving success, and the key to efficient production is tight control of the whole process.

Microelectronics has an ever-increasing role to play in most areas of the manufacturing process. For example, in material-forming operations, microelectronics, in terms of robots, can perform repetitive, boring or possibly hazardous tasks such as paint-spraying operations or welding. Microelectronic controllers are available to accurately perform and monitor machine operations, and computer-controlled (numerically controlled, or NC) machine tools can be used to produce parts repetitively. Microelectronics has also an important part to play in the area of material movements between manufacturing operations and here again robots are used to perform such tasks as warehousing control and assembly line production, notably in the motor car industry.

In addition to these operations, directly involved with the manufacturing process, microcomputers have been applied to the support areas of machine monitoring, where they provide warning of impending machine breakdown or service requirements, and stock control, where an automatic check can be made of stock levels and shortages anticipated throughout the production process. Microcomputers also find application in the product test and quality assurance areas of the manufacturing process, offering more precise measurements plus automatic analysis of test results.

Computers can be used to help the accounting, marketing, distribution, energy monitoring and management functions of the manufacturing process, raising the overall efficiency of the total manufacturing operation. It is apparent, therefore, that microprocessors can be advantageously introduced into a factory at all levels of activity, and wide scope exists for development of this application area.

The word microprocessor will be used constantly here, and so it is worthwhile reviewing the main features of microprocessors with particular reference to those attributes most applicable to process control. A microprocessor is basically a device that continuously performs a sequence of primitive instructions. Essentially the sequence involves determining the required operation to be performed (an instruction), executing that instruction, determining the subsequent instruction and so on. The microprocessor is therefore naturally suited for incorporation as the controlling element for a series of sequential operations – hence its importance in process control and automation.

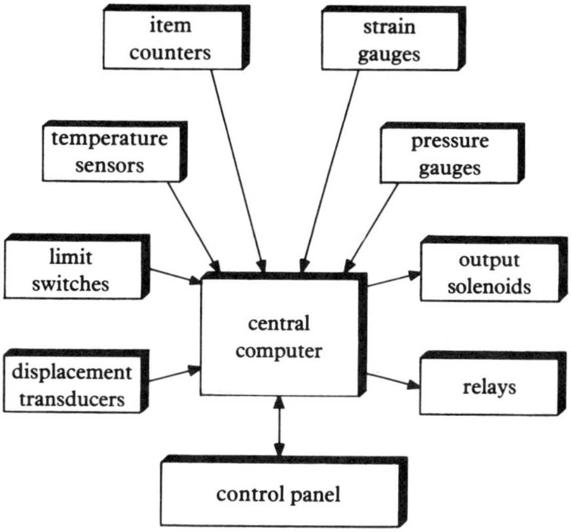

Single controller system. When a single computer is used for the overall control of a process or plant, the computer has to perform several time-consuming tasks. These include monitoring the output voltages from a wide range of transducer types, converting these voltages in an appropriate digital form suitable for computation, correcting for transducer short-comings such as non-linearity and ageing, in addition to performing the actual computations required. For a system with many transducers, these 'housekeeping' operations occupy considerable computer time and the overall response time of the controlling function can become slow.

The so-called microelectronics revolution is not a revolution of ideas but rather the evolution of new components to implement old established concepts in a much cheaper and more compact form. Microcomputers are modelled on the minicomputers introduced during the sixties and seventies and, similarly, the analogue microprocessors that have been recently announced, and which have architectures and instruction sets quite unlike any previous computers, are based on decades of special-purpose digital signal processing practice. Such analogue microprocessors will form the first in a new line of microprocessors tailored in capabilities to the needs of a particular application area. At present, all the available microprocessors have similar architectures and capabilities, although of course some can perform considerably more processing operations per second than others. All microprocessors have basically similar large instruction sets and are electrically connected into systems in an almost identical way. The application most suited to current microprocessors is data processing.

The availability of much cheaper electronic hardware components and, in particular, the arrival of cheap, compact computing power, has led to more and more systems based on programmed logic rather than hardwired circuitry. Indeed, wherever there is a need for computation – no matter how mundane a task – a microprocessor can often be economically employed. In mass-pro-

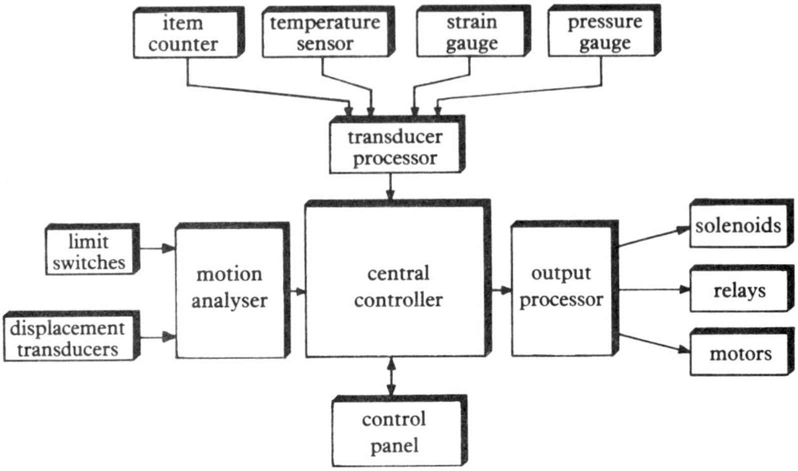

Distributed processor system. When processors are distributed throughout the plant to positions where there exists a need for data manipulation, the central controller is relieved of much of the detailed handling of sensors and actuators. The overall speed of response of the control system is thereby greatly increased, and the system is much easier to build because it has been divided into small, well-defined separate units.

duced goods the incorporation of microprocessors, or single-chip microcomputers, can usually bring very considerable product enhancement, lower costs and a certain measure of self-test servicing. The high product development and tooling costs are spread over huge sales.

At the other end of the applications spectrum are one-off special developments, perhaps to automate the whole, or part, of a unique production line. Here the total costs of the system are little affected by the availability of low-cost microelectronics. The labour-intensive and time-consuming tasks of programming and systems engineering are usually the major cost contributors in such a project. It is under these circumstances that many process-control automation developments are undertaken.

Processing power in the form of minicomputers has been applied for many years to control manufacturing processes. The high cost of these computers meant that they were used sparingly, each servicing a multitude of process sensors, controllers and actuators. The computer spent considerable time collecting the process information and scaling and correcting it for sensor imperfections. The need for so much basic housekeeping processing slowed the response time of the system and also led to detail-laden software, which was difficult to write and subsequently modify. A typical arrangement would involve a large mainframe computer, operating at a managerial level in the manufacturing organisation, being used to supervise the process control minicomputer. However no processing power would exist downstream of the mini, where only transducers, actuators and special interface electronics would be employed.

The availability of microprocessors has led to the spread of cheap computing power down through this arrangement and out to the sensors, controllers and actuators. Smart transducers (sensors with a dedicated microprocessor) send corrected and scaled digital representations of the process variables on demand to the control computer, which is thereby relieved of much of its time-consuming housekeeping activities. Similarly, smart actuators are sent digital demands by the computer which they subsequently translate into the necessary control operations, taking account of all necessary mechanical constraints and any shortcomings in the actuating controller mechanism.

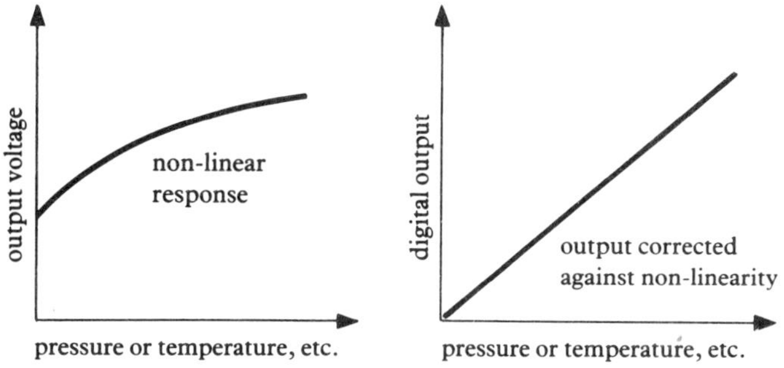

Transducer with microprocessor ('smart' transducer). When a microprocessor is incorporated in a sensor, its output in digital form is compatible with the input needs of the central control computer. This digital output is automatically corrected for zero offset, non-linearity and ageing. Trends, such as rates of change of the measured parameter, can also be calculated and transferred rapidly to the control computer on demand.

The flow of information through an industrial plant can be enhanced in reliability and usefulness by this extension of processing power to the receiving and transmitting stations. One pair of wires circulating around all the data centres can convey and control the flow of all information. Over such a link a computer at managerial level can transmit instructions directly to a supervisory processor at plant level. These instructions can range from simple operation commands to a complete change of the controlling software – thereby enabling rapid major changes to the function of the plant. The increased flexibility thereby achieved in a small batch production environment is especially significant. An additional consideration is that reliable transfer of critical information in a potentially hostile electrical environment is possible only when the transmission and reception are aided by local microelectronic intelligence. The benefits of such distributed processing include the division of the total control and sequencing task into more manageable local assignments such that the system software becomes more structured, leading to fewer mistakes and easier updating. Faster system response is possible, since parallel processing can now take place, and the possibility exists of incorporating local routines to monitor and report on the state of machinery, thus leading to easier (and cheaper) maintenance and higher reliability. System development and system running costs are reduced because of simpler programming and intelligent preventive maintenance, and a more rapid response to changing requirements on the production line is possible.

Since almost all microprocessors operate on information in digital form, there has been a tendency to design process-control equipment based on digital control and this has permitted development of systems that are simpler to adjust, less susceptible to ageing, more able to communicate with each other, and in which self-test and re-programming can be incorporated. The disadvantage is that the personnel dealing with these systems need new skills.

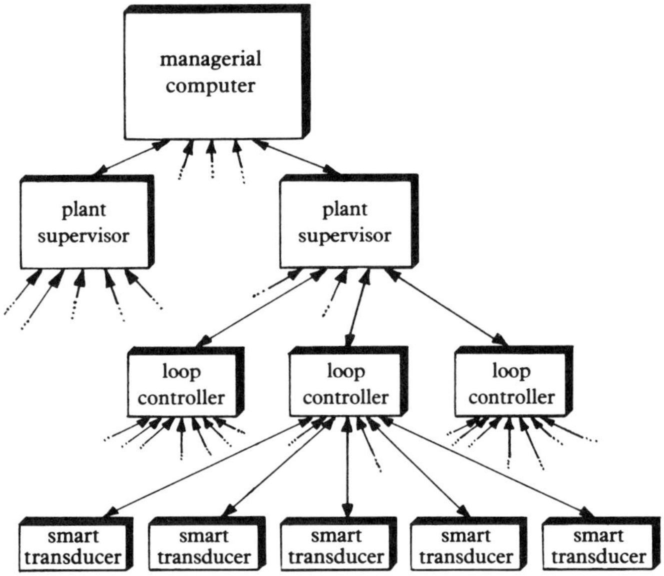

Computer hierarchy. The advent of the microprocessor means that computing power is now so cheap that processors can be economically used wherever there is a need for data manipulation or sequence generation. Computers can therefore be spread throughout a factory to handle tasks ranging from the accounting/management function to linearising the output from a transducer. When all levels of processors are interconnected, the process or plant which is being controlled can be rapidly altered by transferring new programs and data. This leads to very great flexibility in the utilisation of plant.

The subject of microelectronics in process control is clearly of wide-ranging application, but two specific application areas will be considered in depth here, to demonstrate general trends. The specific applications are in production monitoring and control and information gathering. The performance, in terms of output, of a particular piece of machinery can be simply monitored by connecting (interfacing) it to a small electronic package by means of microswitches. Virtually any machine, old or new, can be augmented in this way such that actual production rates for the machine can be measured and compared with a target figure to give an efficiency. Any piece rate or bonus paid to the operator can be automatically calculated and might be displayed to provide incentive. The accumulated data on machine and operator performance can be transmitted to a central administrative computer for incorporation in a managerial database and for payroll calculation. Such a performance monitoring device fitted to a machine would cost a few hundred pounds, and groups of devices can be connected to a master monitoring station situated in the production manager's office.

Although it is possible to build up a process controller tailored for a specific application and using only a microprocessor and several associated components, this approach demands the joint skills of a trained electronics engineer, a computer programmer and a process engineer, and this time-consuming and expensive approach is justified only in highly specialised cases (perhaps where high operating speeds or ultra-high reliability are required). In contrast, a large number of general purpose process controllers, which avoid the penalties of designing from scratch, are currently available commercially. These devices are termed programmable logic controllers (PLC) and available models range from ruggedised versions of microcomputers, with an extended range of input and output modules, to multi-loop digital controllers, which can be programmed by a control engineer with a working knowledge of ladder diagrams or process flow charts. The common component in all the available controllers is a microprocessor. One example, the ITT Director system, costing around $4000, can control several individual sequences and has a programming keyboard that includes the symbols found on process ladder diagrams. It is a relatively simple task to incorporate such a device as a replacement for the relay control logic of an elderly machine. The benefits would be greater reliability plus an opportunity to make the sequence more complex and flexible.

A new Japanese single-loop controller by Hokushin, costing $2000, offers the facility of entering programs via a keyboard or from a tape cassette, and these are preserved in the event of power failure in battery-powered back-up memory. This digital controller has the same front panel controls and indicators as its analogue predecessor and the two types are interchangeable. An eight-loop controller costs about $16,000, and a fourteen-loop system is available at around $25,000. Programming of these devices is simplified by having a menu of control sequences available in memory and which can be employed in the user's programme. A PLC can be readily altered (by programming) to correct, adapt, improve or completely change the controlled process, and if this modification calls for different hardware, then suitable plug-in modules are usually available.

Analogue microprocessors for use in process controllers promise to bring about changes in the design of analogue systems as radical as those introduced to digital systems when digital microprocessors first arrived. Analogue signal processing is such an important feature of process control that the arrival of these devices has important implications for process control automation. Basically, an analogue microprocessor is a device which accepts an analogue signal, converts it internally into a digital representation, processes the data and reconverts it for analogue output. It has an architecture and instruction set tailored to serve this function with particular emphasis on powerful data computation. Digital filtering (previously a task for a minicomputer) can now be carried out by a single chip and the analogue microprocessor is capable of handling signals of frequencies of up to 10 kHz, whereas standard digital microprocessors can cope only with signals up to 100 Hz. Signals faster than 10 kHz require processing by expensive bit-slice microprocessors working in conjunction with very fast multiplier chips.

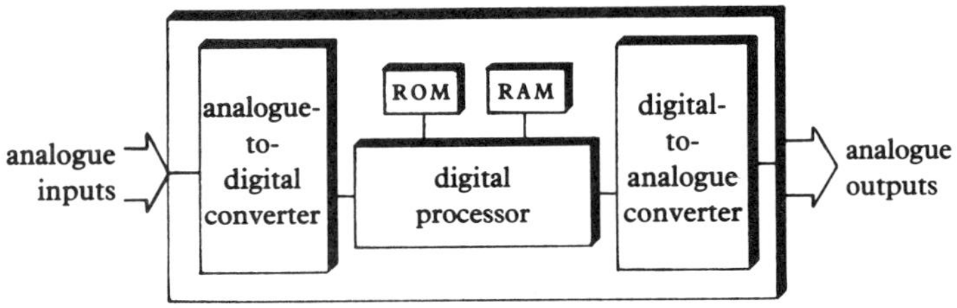

Analogue microprocessors. Advanced microprocessors that have been specially designed for use in process control and signal processing are available. These microprocessors can read voltages directly, can perform high-precision manipulations on this input information and send out the results as voltage signals.

The operation of a controller is limited by the performance of the sensors that actually determine the physical parameters of the process or machine being controlled. Silicon microsensors – transducers fabricated by microelectronics technology, which can be included on the same chip with a microprocessor – are being developed in many laboratories and will soon be used in high-volume consumer applications. The integration of transducers to measure such variables as pressure, temperature, dew-point, humidity, force, acceleration and gas flow, coupled with signal processing on the same chip, will also generate a new class of smart sensors. Sensors may be fabricated as one- or two-dimensional arrays to allow the processing of pressure or temperature fields, and the accuracy of the sensors can be enhanced by periodic calibration, performed automatically. Further, silicon microsensors are around a hundred times more reliable than conventional transducers using discrete transistor circuits.

Computers used for process control can also be used in peripheral roles. High interest rates force manufacturers to keep the level of stocks in factories to a minimum, since material waiting to be machined is dormant capital. By using computers for stock control the level of materials waiting to enter the manufacturing process can be relatively easily reduced. Also, resources (machines and labour) can be scheduled so that there is a minimum of dead time between machining steps. Energy costs can be reduced by a computer-based energy management system. Electricity bills to large users generally have a maximum demand charge, which is based on the actual peak demand, and reflects the costs to the electricity suppliers of installing cables and switchgear to cope with this power level. A special meter can be installed in manufacturing premises to measure the energy consumed over every thirty-minute period and

Peak demand monitor. A peak demand monitor accumulates, during each 30-minute period, the total electrical energy consumed by an industrial plant. If the monitor predicts that by the end of a given period the energy consumption will exceed the maximum level agreed with the supplier, then it can issue alarm signals and shed loads to keep consumption within the agreed limit. The peak demand monitor can be easily installed and can repay its costs of a few thousand pounds within a year in many cases. The display indicates the target consumption as a line and the actual accumulated consumption is displayed as a superimposed bar chart to indicate deviations. *Photograph courtesy of Sension Systems Ltd.*

record the maximum. If the agreed maximum is exceeded the demand charge to the company rises and remains in force for up to three years. Thus there is a strong incentive for the industrial consumer to avoid large peaks in demand, which requires constant, careful monitoring to detect every potential transgression over the agreed maximum demand, and this is an ideal application for a computer. When the computer predicts that a transgression will result if the present power loading is maintained, the computer can shed loads automatically within the plant, or sound an alarm and inform an operator how much load must be shed. The operator can then make the decision as to which loads can be disconnected without endangering safety or product quality. A typical monitor costs a few thousand pounds, is very easy to install and can pay for itself within months in many cases.

Computers also have important roles to play remote from the shop floor. The secretary, accountant, product designer, and indeed all levels of management can use computers to retrieve information, perform calculations, model the product or its market, prepare reports and perform critical path analyses. All of these functions increase the overall efficiency of a company and enable it to react more quickly to changes and disturbances.

Microprocessors are increasingly being used in the design of a new generation of industrial machinery and are used extensively in robots, which have enormous potential for providing safe, continuous, highly repeatable and adaptable operations on a production line. Currently, most robots are single, articulated arms capable of movement under the control of several motors and controlled by a computer. Perhaps because of this human-like feature (the arm) robots symbolise most clearly the potential replacement of human labour by machines and individuals who are unaware of the fundamental limitations of such devices (which have to be told, or programmed, to perform every single step of every operation) can understandably fear that machines may one day dominate humans. In fact a robot is just one more tool invented by man to ease his daily tasks, increase his capability to provide a higher material standard of living and give him greater leisure time. As such the robot is just another step in technological evolution, which started with the wheel. If developments in process-control automation, coupled with advances in automated handling (robotics), automatic inspection and voice input and output to computers, could be confidently applied in large measures to the manufacturing process, the manufacturing potential within a company would be transformed, beyond recognition, by such automation. Of course, the complex process of innovation and technology transfer is not solely technology-driven. Social and economic factors, including the effectiveness of education programmes, the attitudes of organised labour, and the growth rate of the economy will all influence the pace of technological change.

It appears most unlikely that in the immediate future there will be widespread adoption of a bold approach to full automation of the shop floor. Although to follow this bold line would ensure that companies would be in a strong competitive position, it is unlikely to be followed because of lack of investment capital coupled with the inevitable resistance of the workforce, who would certainly be depleted, and the severe lack of implementation engineers.

What is more likely to happen is that new technology will be introduced in a patchy manner, whenever funds allow, in order to improve the performance in certain inefficient areas. In this rather timid process many of the microprocessor-based controllers now in the market offer great potential because of their low cost, their ability to operate independently on local tasks, and their general applicability. In digital process control an expensive minicomputer is now no longer a necessity, so that the entry fee to the digital club is considerably reduced. Companies can slowly extend their application of microelectronic control as and when required and this gradual approach is made possible by the availability of cheap, powerful, local controllers and performance monitors, which have the ability to communicate with other controllers so that complex distributed processing systems may eventually be built up.

8. MICROELECTRONICS AND RELIABILITY

The all-pervasive nature of microelectronics technology means that the reliability of microelectronic devices has relevance to all members of a consumer society that has come to rely heavily on products based on such high-technology devices. Microelectronics plays a critical role in such diverse fields as electrical power generation, computers for aircraft flight control, electronic switching circuits in telephone exchanges and electronic banking systems. The consequences of failure of any of the electronic circuits in such systems will be measured in economic costs, which will be significantly greater than the capital cost of the microelectronics devices used, and in some cases such failure might result in fatalities. The increased complexity demanded by modern microelectronics-based products has created a significant problem in terms of the ability of the microelectronics industry to test these devices for reliable operation. It is often impracticable to perform 100 per cent testing of microelectronic products in a realistic time and therefore such products are largely unproven over the complete range of environmental and user-defined operating conditions. In consequence it is becoming increasingly difficult to guarantee the integrity of such microelectronic-based products without expensive and time-consuming testing.

Reliability is the technical term that describes the ability of a product to operate without failure under prescribed conditions. Failure may be variously described as being sudden or gradual, complete or partial, catastrophic (sudden and complete) or progressive (gradual and partial). The reliability of a product might be expected to be inversely proportional to its complexity, but this is not generally an acceptable situation for equipment in high-volume production where certain minimum standards of reliability must be maintained. Hence, with increasing complexity of products the reliability of individual components must be correspondingly increased.

In the field of microelectronics, new systems and new technologies are generally introduced in advance of the engineering practices that assure their reliability. In such cases it is difficult to predict the contribution to system reliability made by advanced microelectronic devices and, where high reliability is required, the electronics designer can do no more than use his best judgement in selecting components of assured quality and operate them under conservative conditions. The rapid growth of television in the late forties and fifties, and the smooth conversion to colour television, can be attributed directly to the reliability demanded by the electronics industry, where proven reliability of existing equipment is a prerequisite for further growth.

Reliability, which is essentially a measure of quality, is a vital element in each stage of a product manufacturing process, ranging from the initial specification of the product, through design and production, to final delivery to the user. Furthermore, while each of us has a clear notion of the meaning of reliability, its true significance is intimately concerned with the specific product being considered. Thus, in any treatment of reliability it is difficult to be specific, and the material presented here deals with a collection of topics of general importance that, given the attention of the designer, can contribute to improved reliability in electronic equipment.

In designing a microelectronics-based product for improved reliability, several interacting factors must be considered. First, in terms of component reliability, the selection of appropriate component types is essential in equipment designed for high reliability. Furthermore, as the reliability of an electronic component is reduced by operating it at ratings above those recommended, it is essential to design equipment to allow all components to operate within their specified ratings. In many cases the equipment reliability can be increased by component operation well below the recommended ratings. Secondly, with respect to the environmental effects on reliability, components must be provided with suitable physical layout to prevent contact and provide adequate heat-sinking and protection against over-voltage transients and surges. The anticipated physical handling of the product must be considered, as well as the ambient conditions under which the product is to operate. Thirdly, the actual circuit designs will effect reliability, and the circuit must be designed with a view to guaranteeing compatibility of the various component tolerances, at all extremes of operation, and with consideration of the possible effects on other components of a single component failure. The function of every component must be directly related to its proven capabilities as detailed in the published data.

Colour-television design practice, for example, embodies all of these factors. Television sets, which represent probably the single most widespread type of electronic equipment, use a wide variety of components and circuit techniques, are operated by unskilled personnel and must resist many different types of stress. In addition, the operating ambient conditions are uncontrolled and the existence of a competitive market demands high product reliability at low cost. It is of interest to note that the majority of operational failures that occurred in the transition period from black and white to colour television were associated not with electronic components, but rather with mechanical switches and with passive electrical components such as the resistive mains dropper, which is used to provide the different voltage levels required within the television and which was used at the limits of its operating specification in early colour receivers.

The detailed consideration of total equipment design for high reliability is a long, laborious, unglamorous task, calling for great discipline on the part of the designer. A good design represents a combination of science and art, coupled with past experience and careful examination of available information. Codes of good practice have been established over the years and component manufacturers provide comprehensive device information in published data, often supported by applications information. Manufacturers continuously assess the quality and reliability of their products by analysis of the results of tests carried out during and after manufacture. In the short term these results are used to define the quality standards required in production, and to highlight areas where the process could be improved. Information obtained over a longer period, especially from life tests under varying conditions, makes it possible to offer the customer a product of proven and well-defined reliability, and in some cases to offer a range of functionally similar products having varying reliability, with appropriate price differentials. In some cases it is technically feasible to produce components that offer virtually 100 per cent reliability. The price of such components, and of equipment constructed from them, would be prohibitive unless, as is the case with certain defence and aerospace equipment, failure of the equipment cannot be tolerated under any circumstances.

Manufacturers of electronic equipment are, in general, willing to pay the increased costs required to achieve moderate improvements in reliability, because they must consider not only the initial component cost but also the total cost of their product over its useful life. The consumer, choosing between a number of alternative products, must solve a similar problem, normally with much less authoritative guidance. Clearly, the total costs involved in increasing the reliability of a product comprise the initial cost of the components and the cost of field service.

Generally, the reliability of a component is improved when it is operated well within its ratings and less stress is placed upon it. The kinds of stress to which a component is subjected include peak and continuous voltages and currents, ambient and junction temperatures, and duty cycles. Normally the effect of any form of stress is to increase the temperature of the component, thereby accelerating the various physical and chemical mechanisms that lead to failure. Thus, to improve the reliability of a piece of equipment, it is possible for the designer to select components that are more generously rated, achieving operation well within the ratings of these components. The components used are said to be derated, and the increased costs, relative to components that would perform the same function but at the limits of their ratings, constitute the reliability trade-off. The degree of derating applied to components in an equipment can be determined by balancing the increased costs of the derated components against the service costs that would otherwise be required over the useful life of the equipment.

When a microelectronic device ceases to operate properly, the way in which such failure occurs and the reasons for the failure are of major importance to the designer and to the manufacturer. Partial failure does not result in complete loss of function, but causes the component characteristics to deviate from specification limits. Complete failure occurs when characteristics have deviated outside specification limits and there has been a complete loss of function. Failure may be attributable to the inability of the designer to meet specification requirements in such cases as inadequate provision of component cooling, use of the wrong type of components, the unexpected effects of vibration, or the incompatibility of associated equipment causing transient voltages. Similarly, some failures may be attributable to the manufacturing process when the product is subject to stresses within the specified capability.

Typical causes here include wiring faults, incorrect connections, breakages due to poor handling, process faults during plating or welding, and latent faults often initiated by overloading or stressing during check-out tests. A final cause of failure is attributable to misuse, where breakdown can be produced by the application of stresses outside the specified capability of the item or equipment. A typical cause would be the application of weight, voltage, or current overload.

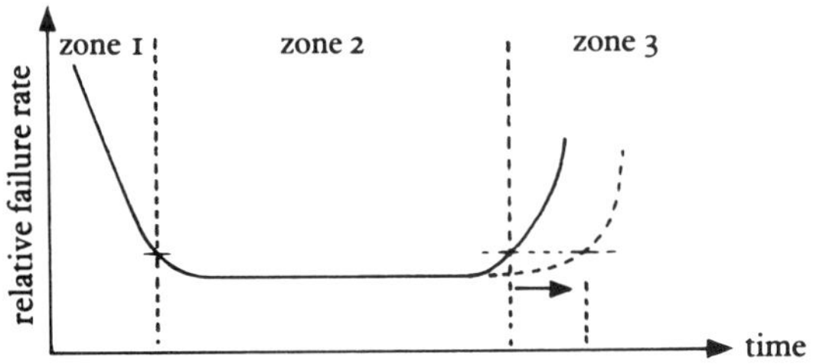

Bath-tub curve of product reliability. The bath-tub curve can be divided into three distinct zones. In zone 1, initial burn-in failures occur due mainly to poor construction, poor quality components or components operating on tolerance limits. This burn-in period should be completed at the place of manufacture. The flat base region of the curve, zone 2 exhibits a constant failure rate and represents the quality of the design and the design life of the product. Zone 2 can be extended by good repair and maintenance procedures (dashed line). Zone 3 represents the inherent wear-out time for the product.

Reliability questions are concerned mainly with expected failure rate over the useful life of the equipment – that is, that period when failures occur in random fashion. It is assumed that early failures resulting from manufacturing errors have been eliminated, and failures due to the equipment wearing out have not yet begun because of the short time during which the product has been used. The failure rate of a product demonstrates a 'bath-tub' characteristic, with high failure rates in the early and late stages of the product's life.

In the first period of product life (typically the first 100 hours) a relatively high rate of initial burn-in failures will be produced, mainly due to poor construction or to components that are being operated on tolerance limits. This burn-in period should ideally be completed at the place of manufacture, since failures in this period are the direct responsibility of the manufacturer. A more detailed analysis of the failures in this period of the product's life shows a distinct subdivision between failures due to faulty proprietary items (electronic components) and failures due to faulty equipment design and manufacture. The ratio of these two failure sources will depend on the type of equipment. Use of components that have suffered deterioration in storage or transit, or that have been damaged during the manufacturing operation, normally results in failures that appear as soon as the equipment is tested. This type of failure is described as infant mortality, and is unconnected with the user's operation of the equipment. A computer using a large number of components in well-designed, repetitive circuits might give a ratio of failures due to faulty components to failures due to faulty manufacture of about 4:1. In equipment such as communications receivers or guided weapons this ratio might well be reversed, because design and manufacturing problems are greater owing to the employment of advanced devices and other sophisticated techniques. Failures that appear repeatedly in several individual samples of a product can be easily diagnosed and corrected by the introduction of modifications or changes in the appropriate production procedures. In contrast, failures that show different characteristics in separate samples of a product but that manifest themselves in the early stages of operation are generally due to imperfections in design or construction.

For the main period of the product's life the rate of failure of the product will show a random distribution. The actual failure rate is a measure of the quality of design and the design life of the equipment – the lower the failure rate the better the reliability of the equipment. During the early part of this period the product is normally covered by the manufacturer's warranty agreement. For domestic electronic equipment it is important from the manufacturer's point of view that a minimum number of failures should occur during this time for two main reasons. First, included in the purchase price of any product is a charge calculated from a statistical estimate of potential failures and the associated cost of repairs during this warranty period. If the failure rate is higher than anticipated, financial losses will follow and the product may cease to be a viable proposition for the manufacturer. Secondly, poor reliability during the warranty period will lead to consumer opposition to the product, with an attendant reduction in sales. The life of the equipment can be extended by good repair and maintenance, such that the useful working life of the product when only such random failures are present will be as long as possible.

Failures during operational life can often be reduced by redundancy techniques, which can be relatively simple to use. For example, in cable looms it is often the case that multi-pin connectors have a few spare pins, allowing the most important leads to be used in parallel. In addition to use of such redundancy, fail-safe designs using fuses or circuit breakers can be used to reduce damage. Short circuits during testing are a common cause of damage to power supplies or semiconductors and can be prevented by good design. Safety circuits can be designed to revert to a safe state if voltage is lost and in weapons, such as missiles, memory circuits can be employed to keep the missile on course if the target is temporarily lost.

Failures that occur towards the end of a product's life are time-dependent and are often caused by mechanical and chemical action. Most electromechanical parts tend to wear, and all the designer can do is to minimise the rate of wear by careful choice of materials, employing sealed units to protect components from moisture and atmosphere, using derated components and avoiding high temperatures. Preventive maintenance will also defer the end of life until the point where service costs become excessive.

To give a feel for the quantities involved in electronic equipment reliability, consider the case of a piece of equipment that incorporates a number of microelectronics devices whose individual failure rate is 0.1 per cent per 1000 hours, and where the design of the equipment is such that if one component fails the equipment will fail. If the equipment incorporates only ten components the probability of the overall equipment operating without failure over 1000 hours is 0.991. If the equipment contains 700 such components the probability of operation without failure reduces to 0.50 over the same 1000 hours.

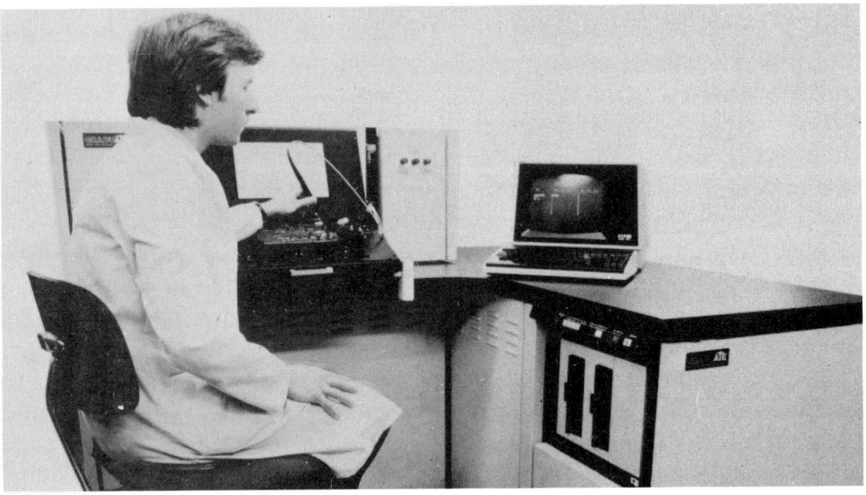

Automatic test equipment (ATE). Modern electronic test equipment exploits a principle known as 'virtual instrumentation' such that the circuits under test 'think' that they are operating with standard conventional test instruments when in fact all previously required test instruments have been replaced by software-controlled circuits. Thus knobs, dials, displays and meters have been eliminated and the system can be quickly reprogrammed to perform a variety of tests on a variety of circuits. *Photograph courtesy of Marconi Instruments Ltd.*

Often, a failure in one device can be caused by a fault in another part of the system. This is a systems engineering problem, and is often due to the fact that different parts of the system have involved different levels of development such that all of the possible fault interactions cannot be fully anticipated by the designer. This raises the problem of human error, which may be one of bad judgement, lack of experience, carelessness, or inability to make decisions quickly enough. By good systems organisation, adequate documentation, procedures and checks, the effects of human error can be minimised, but it will still be a cause of unreliability and will tend to increase in significance as the complexity of the system increases. The implication that this statement holds for the designers of microelectronic equipment employing complex integrated circuits has not yet been fully recognised. However, it is clear that such designers and manufacturers must be constrained to produce self-testing or, at least, easily-testable circuits.

One of the most basic tenets of reliability theory is that the overall reliability of a system that consists of a series connection of elements is less than or equal to the reliability of the least reliable element, whereas the reliability of a system consisting only of parallel elements is greater than or equal to the reliability of the most reliable element. This result leads to the use of redundancy techniques in product designs.

The dramatic increase in capability and availability of LSI microelectronic circuits has only recently made redundancy a practical proposition. Digital systems can now be designed to be continuously and totally self-checking such that it is possible to locate a failed module when a fault condition occurs. In some cases a ten-fold increase in product life can be achieved by only a three-fold increase in circuit complexity. It is interesting to note that self-checking digital systems can now be designed to obviate the need for highly trained operators and maintenance personnel. For example, in a memory arrangement with one or more spare memory modules, hardware and/or software checks can be made to ensure correct operation. If failure is indicated the appropriate unit can be deactivated and a healthy spare can be assigned to take over the affected part of the memory. The response of the designer to the increasing complexity of integrated circuits has been to build testability into a circuit at an early stage and, more significantly, to make extensive use of redundancy and error correcting codes. This change of approach will of course change the nature of the testing methods used and in some cases could obviate the need for detailed chip testing.

As future microelectronics-based products will increasingly rely on circuits of higher and higher complexity, designers of such LSI and VLSI circuits must address themselves to the problem of reliable circuit operation, and must become aware at the circuit design stage of the need to produce a self-testing product or, at the very least, easily-testable circuits. Assuming that design errors have been eliminated, initial device failures may occur as a result of fabrication problems such as mask flaws, dust particles and defects in the silicon crystal structure. Approximately 60 per cent of all manufacturer's rejects can be attributed to package, metalisation or bonding defects. Surprisingly, 80 per cent of in-service device failures can be attributed to misuses such as test-induced faults and electrical overstress after applying incorrect drive voltages, or to wrong diagnosis such as predicting an integrated circuit failure when in fact the failure was due to other circuit faults. The remaining 20 per cent of device faults are normally caused by packaging failures and the delayed effect of wafer fabrication defects. Often a batch of devices will operate satisfactorily under normal ambient conditions but will progressively degrade and ultimately fail when operated at the limit of the device operating range. Charge accumulation and current-induced movement of metal atoms are the principal causes of failures that can be induced by operating at the extremes of device operating conditions. The migration of metal atoms is the most significant long-term degradation mechanism in semiconductor devices, where at high current densities and elevated temperatures metal atoms are attracted towards the positive end of a conductor and metal disappears from some critical region until ultimately an open-circuit and therefore failure occurs.

These progressive failure mechanisms are aggravated by operation at high temperature and for MOS circuits it has been found that failure rates will double

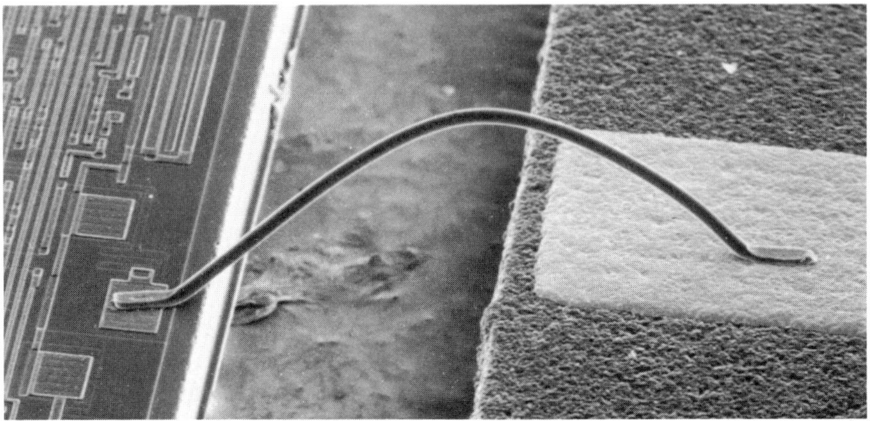

Scanning electron microscope. The scanning electron microscope (SEM), which offers magnifications greater than 100 times those achievable with conventional optical microscopes, forms an invaluable tool in the characterisation of integrated circuits. With SEM analysis it is possible to detect minute fabrication defects such as cracks in metal connections or surface impurities which would affect the ultimate reliability and lifetime of the device. Advances in SEM products now permit the actual operation of the device under test to be monitored by a technique which displays device voltage information as colour contrasts on a colour television screen. This SEM photograph shows a bonding wire 25 microns thick connecting a metal pad on the integrated circuit chip (left) to another metal pad on the package (right).

for each 10 degrees C temperature rise. Local temperature increases caused by individual transistors operating beyond design limits can be a much more insidious cause of trouble. Such temperature increases may not immediately impair the overall operation of an integrated circuit but may accelerate one of the progressive failure modes. This is a difficult situation to investigate, because thermal time constants are much longer than the testing time that can be economically allowed for each circuit. Fortunately microelectronic devices can be fabricated to be so reliable that, under normal operating conditions, years of testing would be required to predict their reliability. A typical rate of 0.1 per cent per 1000 hours is equivalent to only one error, on average, per 1000 units continuously tested for six weeks. The greater the reliability of a device the more difficult it is to determine its reliability. Curiously, the temperature dependence of the failure rate can be used advantageously to test devices in a reasonably short time. When a device is run at a higher than normal temperature a much higher failure rate will result. Results obtained at more severe stress levels can then be extrapolated to normal stress levels to give an estimate of the life distribution. Such tests are called accelerated-stress life tests.

The large increase in the levels of complexity that can be utilised in a silicon integrated circuit is causing designers to consider testability at a very early stage during a product development. A logic circuit having sixteen inputs will require approximately 60,000 different input patterns to test all possibili-

Probe testing of integrated circuits. Integrated circuits are tested before they are individually sliced from the silicon wafer. Processed wafers are held on a controlled platform by means of suction and an array of fine test probes is used to make contact with specified points in the circuit structure. When the probe tips meet the wafer surface testing is performed automatically, row-by-row, until each individual integrated circuit has been tested. Faulty devices are marked by an ink dot and are subsequently rejected, so that the reliability of integrated circuits is monitored from the early stages of manufacture.

ties. If the circuit contains memory elements whose output depends on previous input and output states, then the combination of internal states with external inputs must be considered, and it very rapidly becomes impractical to test all possibilities in a sufficiently short time, and a long testing time may lead to a product that is too expensive to produce. As a consequence of this testing difficulty there is a very real possibility that malfunctions will occur in complex devices when they are put into a particular state since this may be the first time that the device has operated under these specific conditions. A particularly difficult situation arises when failures are pattern sensitive. For example, a memory cell may be able to store 1 and 0 when its neighbours are 0 but may be held permanently at 1 when its neighbours are 1. Testability can be improved by partitioning into simple units, providing access to internal states and by using self-test techniques. Of course extra silicon area must be used but as a by-product of the progression to larger (VLSI) circuits the additional space required is often less than 10 per cent of the total chip area.

Error-correcting codes such as Hamming codes can be employed to increase the reliability of a circuit. Basically, additional bits are added to each data word with the objective of creating a new set of data words, each one separated from the others by a big enough binary difference to enable errors to be detected and corrected. Such error-correcting techniques have been extensively used in memory devices and have been found to be particularly successful because 90 per cent of memory operational failures only involve a single bit. In fact single errors can be corrected and double errors detected by the Hamming code method.

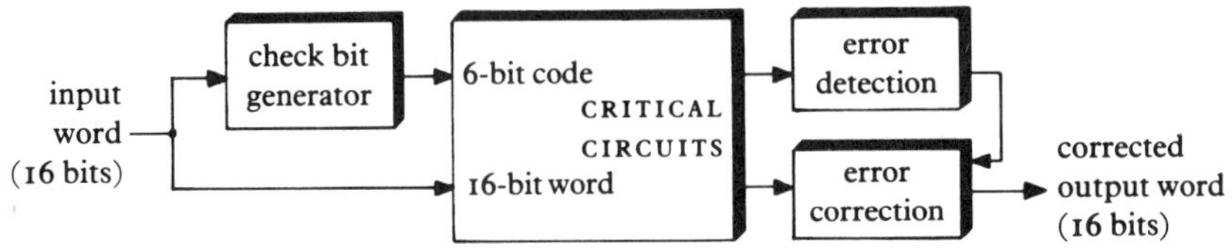

Error-correcting circuitry for high reliability. In a microelectronics-based product that requires high reliability, error-correcting codes (Hamming codes) can be employed to detect and correct errors. For a 16-bit processor an additional 6 check bits are required. The outputs from the processor can be checked for errors and the extra check bits can be used to correct these errors automatically.

If a chip using error correction is tested with the error corrector disabled, chips of the highest reliability will be produced because the entire capability of the code is then available to correct for failures that occur after testing. However, by testing with the error correction operational, a much higher percentage of acceptable (working) chips will be obtained. The increased yield of acceptable chips over the yield of perfectly tested chips must be paid for by a decrease in reliability. It has been found that the penalty for obtaining a significant yield increase by this method is not severe. Indeed with adequate partitioning, circuit redundancy and error-correction circuitry included in the design of an integrated circuit, the potential exists for the production of high-reliability microelectronics-based products that require no production testing.

Flexibility is a very common design requirement for complex control and monitoring systems and is typically achieved by the use of stored program techniques. Consequently, reliability and maintainability characteristics of complex software packages must be considered in addition to the reliability of the hardware on which the software is mounted. Hardware design errors approach zero as a function of time because designs must be fully tested before production commences. However, it is a common experience to find that software design errors decrease with time to some value that is a function of the rate at which new capabilities are added to the software. Software failure is not due to a wearing-out process but usually occurs only when a program is exposed to a problem for which it was not designed or tested. High reliability can only be achieved at high cost, and in most cases it is probably not cost-effective to remove the last error that, if not corrected, would cause only a trivial failure once in a few years of continuous execution. The size of the software module has a dramatic effect on the design and verification effort necessary to assure reliability in large systems. More work is required to define design procedures that lead to software partitioned into easily testable modules and subsystems, but perfect software reliability is unlikely to be achieved.

A change in the traditional man-machine interface is a further consequence of the increasing complexity made possible by microelectronics and software techniques. It is not clear how intimately the human operator should be involved in the control of complex systems. For example, timely intervention of an operator can avert catastrophes where automatic recovery procedures would have failed. Conversely service degradations or, in the extreme case, system failure or crashes can be caused by erroneous human actions. Clearly the designer concerned with reliability must globally consider hardware, software and the man-machine interface.

The problem of reliability must be tackled differently depending upon the specific application area. For example, electronic equipment designed for use in manned spacecraft must offer a high confidence level of survival in space for a long period and must also be able to withstand the extreme conditions experienced on take-off and re-entry. Paradoxically, an aircraft landing system is required only to be reliable during the minutes of actual touch-down but with a high degree of repeatability. Nevertheless the need to enhance reliability, however measured, is common to all microelectronic devices and all microelectronic-based products, and the reliability achievable with future microelectronics developments may yet prove to be one of the principal factors limiting the widespread application of microelectronic technology.

9. MICROELECTRONICS AND ITS IMPACT ON SOCIETY

Man has repeatedly demonstrated a healthy suspicion of new inventions and has been generally reticent about accepting new ideas. Recent examples such as the steam engine, the motor car and the aircraft come immediately to mind but the basic tenets of modern science, developed in the nineteenth century, were equally revolutionary and at first generally unacceptable. It was, however, in the nineteenth century that the idea of progress in society through science took shape and the connection of progress with science has been fundamental to the way in which the Western World has developed ever since. The development of machines through the nineteenth century resulted in a very different society from that based on agriculture, which it replaced. Initially there were objections to the introduction of new machinery and the Luddites are well known, even today, for their physical attempts to prevent progress. Their protests, however, did not succeed and after only one or two generations there was general acceptance of the new machinery. The reason was that, in spite of the complete change involved in the type of work and the way in which work was organised in factories, mechanisation produced for very many people new levels of personal wealth. The scientific triumphs of this period were displayed at the Crystal Palace Exhibition of 1851, an exhibition that suitably demonstrated Britain's leading position in the development of technology at the time.

The reason for returning to the first industrial revolution is not simply to remind ourselves of the former greatness of Britain, but to draw a parallel with the new technology of the twentieth century, microelectronics. Once again man is faced with the prospect of dramatic changes in his working pattern yet the potential for progress is great. The word potential is used here with some care, for although microelectronics has been around for a number of years and has found widespread application, only the tip of the iceberg is so far evident and below the surface lies the real potential, which will be realised over the next decade and beyond. During the next fifty years, microelectronics will pervade almost every aspect of society. Just as, in the first industrial revolution, machinery was used to overcome the limitations of the human physique, so in the second industrial revolution, as it has been called, microelectronics will be applied to overcome the limitations of the human brain. Such a prospect is indeed awesome, and it is not surprising that many people are apprehensive about the development and impact of microelectronics. Is this apprehension simply fear of the unknown, as with the introduction of mechanisation, or is there something qualitatively different about this particular development? The previous chapters of this book have discussed the technical basis of microelectronics in some detail. With this knowledge it is possible here to consider the effects that the new technology will have on society.

The development in the twenties and thirties of radio and television produced the first widespread impact of microelectronics on society. As radio and television were directed towards leisure and improving communications, they produced little threat to existing industrial practices and were accepted fairly rapidly and without much social discontent. Ironically, it was the depression of the thirties that helped to create a demand for the new, low cost, broadcast entertainment to provide escape from the difficulties of the time. Today, it is the development of the computer and its microelectronic equivalent, the microprocessor, that is providing the greatest challenge to society. The ready availability – at very low cost – of computing power is allowing many new products to be developed and is making possible a high level of automation, which involves replacement of the decision-making role of a human operator by an electronic circuit. As it is difficult to grasp easily the consequences of having the power of a computer available as a single component, consider a few examples of products based on such devices. Essentially what is happening is that products are becoming smart, they are able to make intelligent decisions, perform calculations and memorise results for future use, in addition to their usual functions. A simple example is the multi-function electronic watch, which, in addition to giving the time of day, will also calculate correctly the date and day of the week, making the necessary allowances for different months and leap years, will allow accurate stopwatch timing of events and even produce an alarm when a pre-programmed important date such as a wedding anniversary is approaching. The most elaborate watches include complete scientific calculators in addition to the full range of timing functions. As watches are one of the major common uses of microelectronics, developing new uses for the computing power of the watch is currently a major industry. Recent ideas have been to include medical functions, such as heart rate monitors or pulse rate indicators. Special sensor circuity is required to detect the physiological parameters, which are transformed into digital signals for the microcomputer to do the necessary calculations. A direct reading of the pulse rate is of great interest to athletes, stressed businessmen, and the medical practitioner. These are only the first steps of development. In future, other physiological parameters will be monitored but, more importantly, the watch may take on the role of the telephone handset and provide a more convenient remote communications system.

With existing technology the telephone handset provides a second example of how intelligence is being incorporated into products. The recently introduced electronic payphone provides modern push-button dialling with a digital display of the charges as the call proceeds and, unlike the previous traditional payphones, a variety of coins can be used. This provides a much improved service to customers but, perhaps more importantly, the microprocessor provides another feature that is not directly obvious to the user. The electronic controls provide a built-in fault-finding capability to simplify maintenance of the unit. This is a most important development, which is beginning to be included in the majority of microelectronics-controlled products. It is almost trivial, having decided to use a microprocessor in a product, to use it to detect when something goes wrong and to pinpoint the fault when the instrument is tested by the maintenance engineer. This capability is revolutionising the maintenance requirements of increasingly complex equipment, which otherwise would require the widespread use of highly skilled design engineers to repair faults. Looking one stage further it is now considered feasible, because of the low cost of microelectronic devices, to provide significant redundancy in the design of a system so that if a fault occurs in one section of circuitry, the signals are channelled through standby circuitry.

Computer chess. Computer chess programs which attempt to simulate how human players think and anticipate future moves have been developed over the last thirty years. The most recent product incorporates microelectronic circuits to produce a voice output to speak every move that the computer makes. In contrast to previous models where each move was required to be keyed-in with a keyboard as well as manually playing the chess pieces, this latest model has each move entered automatically into memory. Games of different skill levels are available as are pre-programmed games by leading grandmasters against which the player can try his skills. *Photograph courtesy of Computer Games Ltd.*

Electronic games and toys provide a good illustration of the power of the microprocessor and the smartness that can be added to a product. When the first electronic games were produced they used simple fixed-logic circuits and hence were only capable of giving one or two types of game and these were of limited interest. With the introduction of the microprocessor it became possible for the various facilities of the game to be altered at will. Not only can a large range of games be provided (some systems provide over 200 different games), they can also be much more involved, and the degree of difficulty for one or more players can be progressively increased as the players become more skilled. One particular example, optimised in this case for the game of chess, which can be played at different levels of difficulty, demonstrates the sophistication of the type of game which can now be achieved. It is claimed that the most difficult level of one of these games, the 'Chess Challenger', which is built around a fairly unsophisticated Japanese microprocessor, is equivalent to a very good club player!

Electronic language translator. The electronic language translator allows words and phrases from one language to be translated into another and displayed or spoken to the user. Complete words can be keyed in or, by means of the numbered keys, a selection of entire sentences can be translated directly. Available products contain a vocabulary of over 700 words and phrases but as technology improves this number will increase substantially. *Photograph courtesy of Texas Instruments Ltd.*

Perhaps one of the most interesting smart products to be announced is the electronic phrase book. Complete with its own voice, this product allows the user to key in the word or phrase to be translated and almost immediately the answer will be displayed on a small screen and, if necessary, spoken, with an acceptable accent. Several languages are available, including Japanese. At the present time the vocabulary is restricted by technological limitations to about

200 words, but if the expected progress is made, it will only be a few years before complete dictionaries can be stored in hand-held units. The final stage of development will occur when the input can also be spoken instead of typed by means of a keyboard. Automatic speech recognition is, however, a more difficult task than speech generation by computer, and it seems unlikely that voice input will be commonplace before the end of the century. The impact of this development will of course be much wider than just electronic phrase books. The whole concept of computers will change when the barrier of the keyboard input is removed.

These few simple examples perhaps begin to put the capability of microelectronic devices into perspective. The number of smart machines in existence today is still relatively small, but it is more than is commonly realised. Microprocessors are now used extensively in sewing machines, dish washers, washing machines, telephones, cookers and ovens, controllers for heating systems, televisions, cameras, traffic lights, car ignition systems, taxi meters, lifts, security systems, industrial processes, energy management systems, offices, accounting systems, cash terminals, banking terminals, library indexing systems and many more.

To a large extent microprocessor and microcomputer chips have, until now, been made by the semiconductor manufacturing companies and it has taken some time for applications to develop and for the products to evolve. The enormous investment in the technology, however, is now paying off and many products are appearing. Such is the rate of growth, indeed, that the semiconductor companies now are finding it difficult to keep up with the demand. However, new product developments are limited either by the generation of good ideas or by the supply of suitably trained people with the skills to implement them.

Some particular sectors of the microelectronics market are experiencing growth of 50 per cent per annum and it is interesting to note that the semiconductor industry, which only came into being in 1950 and produced less than $200 million worth of integrated circuits as recently as 1956, is expected to produce in excess of $15 billion worth of circuits by 1985. This, of course, only relates to one sector of components, and if one looks at the overall electronic market the total turnover in 1985 will exceed $800 billion, making it one of the largest industries and wealth-producing activities in the world.

How then does this new industry and the microelectronics revolution affect society? With something as all-pervasive as microelectronics it is almost axiomatic that there will be changes directly attributable to its introduction, and whether or not these changes can be considered beneficial will depend entirely on one's point of view. Consider first the most fundamental use of microelectronics, in the computer. Large computers have been around for many years and have been used successfully for processing large volumes of information such as that associated with National Insurance contributions, taxation, salary payments, banking records, staff of companies and Government departments. Although the introduction of these large machines has removed some of the more traditional clerical activities, different jobs have emerged in connection with the preparation and input of data to the computer.

Such computer installations were, until the recent microprocessor developments, restricted in their use to large organisations because of their cost, typically greater than \$1 million. A second important feature was that the various systems were independent and it was not practical to move information from one computer to another. This provided a certain degree of security to the information, because personal details could only be accessed by the operators at a particular computer, and it was unlikely that the information would be used for some unauthorised purpose. This does not, of course, prevent bodies building up files of detailed information about citizens, which may or may not be correct and which cannot be subsequently verified by the persons involved. This question of data banks containing personal information is of fundamental importance and is hotly debated, particularly by the Council for Civil Liberties. Even in the United States, where there is greater protection for the rights of individuals, the problem of personal information held on computer files is not fully resolved although there appears to be a greater awareness than in this country that a problem does exist.

The recent developments of microelectronics, however, make the situation potentially more acute, since not only does the improved technology reduce the price of computer systems so that a wider range of people can afford them, but it also allows a parallel development of communications technology so that computers can now be linked together in a network, with information being freely transferred from one computer to another. This has many obvious benefits, but, in the above context, personal information gathered for one purpose can now be accessed, at least in principle, by a second party who can pass it on to a third party and so forth. The technical capability already exists for extensive networking of computers and networking is in widespread operation for scientific purposes. Society should consider carefully whether or not it is satisfied that there are sufficient safeguards built into the systems to ensure that, in the interests of administrative efficiency, personal freedom is not at risk.

Large computer systems will continue to be used for large-scale data-processing requirements, just as they will for the sophisticated calculation of scientific research. However, for many areas, where a few years ago centralised mainframe computers would have been required, the new breed of desk-top microcomputer systems are taking over. Although these microcomputer systems do not compare with some of the capabilities of their larger brothers, the difference lies mainly in the amount of memory that they contain and the fact that they can only do one operation at a time, in contrast to the large computers, which will do many tasks simultaneousy. However, the cost benefits of the new system tend to outweigh these disadvantages. The small microcomputers can be considered as personal computers and in the United States, where technology is more accepted, they are sold to householders as home computers for the management of the family affairs. With some additional electronics known as interfaces they can control the heating system, simulate occupancy by switching on lighting at random to deter burglars, control telephone-answering machines and act as word processors. In this country, where we are either more conservative or less wealthy or both, this aspect of the microcomputer market has been mainly limited to the computer hobbyist. These microcomputer systems are, however, providing an interesting alternative to the mainframe system for the small business. At a cost of a few thousand pounds nearly every small business can afford a system able to help with a wide range of functions, including accounting, secretarial duties, stock control and tax returns. The most obvious area of application is in the office as a word processor. In spite of the impressive title and the impressive performance of modern word-processing systems, they are not very demanding on the technology. The critical parts are the printer and the keyboard (the mechanical parts) whilst the microprocessor and memory is able to manipulate the text with ease. Microprocessor systems are also ideal for keeping records, although on a much smaller scale than mainframe systems. One interesting example of this is in a medical practice, where several thousand confidential patient records have to be kept. This amount of information is easily handled by a microcomputer with floppy discs to store the information. Before a patient is interviewed by a doctor, he can call up the information on the patient and his previous treatment from the floppy disc by typing in the patient's name at a keyboard, and the information will be displayed on a screen. During the interview, details of the consultation and any medicines prescribed can be added to the computer file on the floppy disc, and the information on the screen deleted. As the records remain in the surgery, they are at least as secure as the existing card files. Such a system, when programmed properly, will provide features not at present possible. The generation of prescriptions for patients on long-term treatment can be automated, for instance, and in a more sophisticated system it is possible for the doctor to be helped with diagnosis by feeding into the computer a set of symptoms.

The application of microcomputers is still at a very early stage. In most cases the software programmes that control the computer are not available even for some basic accountancy functions. This highlights an important point often overlooked when discussing microcomputers – the software. Whilst the glamour is attached to the technology and the hardware developments, the application of the devices and systems require considerable software to make them operate in the desired way. The semiconductor manufacturers have overcome the technical and manufacturing problems of producing standard circuits at low cost for use in a wide range of applications, but the consequence of this is that the programming of the product is left to the system manufacturer, and is a cost that he has to bear directly. The added value in many products is now made in terms of the software developments.

The overall effect of all these developments is that microcomputers will become everyday tools of business and professional activities. It should be noted that it is very often the professional person who uses the system directly. Because of the way in which information can be accessed directly at a terminal in response to keyboard instructions, the requirements for a secretary to retrieve records or to file papers is removed and there is obviously a possibility that many clerical jobs may disappear. The corollary of this is that individuals will require a familiarity with computers at least at user level, and, if they are to make full use of them, at a programming level too. This will prove problematic in the short term, and one might see an explosion of courses in computing for businessmen, but for the longer term it is clear that familiarity with computers must become a basic school subject.

The appearance in the educational system of computers, and microelectronic technology in general is, of course, fundamental to its long-term development. Many schools currently have microcomputer systems and a number of projects are under way that aim to make the subject much more widely appreciated in schools. A measure of the importance attached to the subject can be gauged by the fact that formal school examinations in Computer Science are beginning to appear.

Apart from the teaching of computer techniques as a curriculum subject, an interesting question is just how far the technology will feature in the teaching process itself. Microelectronics has become widely used for programmed learning exercises and will be used quite generally for a wide range of subjects. There will, of course, always be a requirement for the teacher, who supplies much more than just information, but the ability of smart microprocessor-based learning stations to present information in novel, entertaining and challenging ways will feature to an increasing degree. As the cost of hardware falls, one of the present serious obstacles will be removed and one might expect a rapid introduction of microcomputer systems on a time scale dictated principally by the massive task of the development of all the necessary software.

To return to the concept of the convergence of computing and communications. We do not have a simple word for this, but the French have dubbed it 'telematique' in the important Nora Report, which provided a blueprint for the future development of the French communications systems. The French were, in a sense, fortunate, because their telecommunications system had been allowed to decay to such an extent that it was amusingly typified by the aphorism 'Half of France is waiting for a telephone. The other half is waiting for a dialling tone'. As is often the case, it is easier, at least politically, to replace a totally obsolete system than make improvements to an almost acceptable one, such as we have in the United Kingdom. Consequently the French are demonstrating one important way in which microelectronic technology can be used, having decided to provide everyone in the country with an intelligent computer terminal that will be linked, by means of the new telephone network, not only to all other terminals but also to centralised computers. The first reason for doing this is to eliminate the printed telephone directory, which the French maintain can be readily stored in the computer and easily kept up to date. The French Post Office calculate that they can afford to give user terminals to subscribers free of charge as they will cost no more than the printing and distribution costs of the directories over a five-year period.

The home visual display unit, linked by the telephone system to a computer and to other VDUs, opens up a number of possibilities, some of which are actually operational in this country, although the terminals, which are fairly expensive, have to be purchased separately. The information on the central computer does not need to be limited to telephone numbers and an extensive information bank has already been established by British Telecom. The system, known as 'Prestel', is based on a regional network of computers and is used by a wide range of British industry. Some pages involve advertisements, suppliers' information, detailed stockmarket reports and financial information, travel timetables, aircraft seat availability, etc. The service has only been running for a short time but already several hundred pages of information are provided and as more receivers are purchased the range of topics covered will expand. It is probable that this form of information service will create an increasing awareness among the population and have an effect on patterns of shopping, purchase of books, use of libraries and selection of holidays.

Prestel. The Prestel data service available from British Telecom uses the standard telephone system for data transfer. By purchase of a special 'modem' interface, it is possible to display a range of information on a normal television screen. Prestel facilities include local 'What's On' information, news, weather, recipes and can even allow seat bookings for theatres, etc. *Photograph courtesy of British Telecom.*

The Prestel system is used primarily, at present, for receiving visual information but, with suitable terminal equipment typewritten letters could be transmitted over the telephone system. Such systems are already in use in business. Further of course, the 'letter' does not need to exist as an actual piece of paper but could be displayed on the VDU for verification and then stored on a floppy disc. Financial transactions can be undertaken using this technology, debiting the bank account of a shopper and crediting the account of a supplier or retailer, thereby easing the physical requirements of paying bills, if not the financial ones. This may seem a long way in the future, but steps along these lines have already taken place with the 'Cashline' and 'Autoteller' banking terminals. With these, the user communicates directly with a mainframe computer at the Bank's head office to adjust the account according to the amount withdrawn. With the home terminal, it will not be possible to withdraw cash but, because of the increased privacy, it will be possible to carry out more extensive transactions such as transferring funds from an account to a

Electronic funds transfer. Electronic cash terminals, which offer cash facilities 24 hours every day including weekends and Bank Holidays, are now positioned in all major U.K. towns and cities. The microprocessor control cash terminal is in direct communication with the central database of the specific bank and can provide up-to-date account information as well as cash. These cash terminals operate only when used with a personal credit card-like identification card and a personal code number known to the individual. *Photograph courtesy of the Royal Bank of Scotland Ltd.*

range of suppliers and the need for physical money will be reduced. This is the beginning of the cashless society.

Such information transfer will not be limited to the home terminal. The old-fashioned cash register, which has been automated to calculate the correct change and to provide data for the stock control system and is now called a Point-of-Sale terminal, can just as easily be connected to a banking computer network. The amount of the bill could then be directly debited from an account rather than a cheque being written or a credit card being used. Objections to such direct debiting systems have been raised. One particular problem is the question of what happens when the computer tells the cashier that the customer does not have enough money in the account to pay the bill. This would certainly be embarrassing, but it could be considered better than being unexpectedly overdrawn and the Bank not accepting the cheque. It is certain that acceptance will be gradual, just as it has been with credit cards in this country, but it seems that the increased convenience for a significant sector of the population will be sufficient justification for electronic fund transfer to be adopted, perhaps in parallel with normal cash terminals.

These examples illustrate that through the convergence of computers and communications the storage and transfer of information will be dramatically improved. Consideration of the methods of storage and communications will no longer be a preoccupation, they will be taken for granted, they will be microelectronic. Filing systems, mail services, reference information systems, telephone and telegraph systems will all become integrated and the focus will be on the important issue – the information itself. Hence the phrase that has been coined for our future state, the Information Society. This will not, of course, happen overnight. It will grow gradually and it will grow at different rates in different countries. Britain, surprisingly, is one of the front runners in developing the necessary communication links with the combined voice/data telephone network due for operation in the near future. The major concern here must be the question of security of the information to be transmitted and the protection of the individual. This is not a trivial matter and satisfactory safeguards should be established before widespread use of the system is accepted. Data protection legislation has been introduced in the United States and several European countries. The levels of protection vary but all provide some form of right-to-know so that an individual has the right to find out what information is on record about himself. In Britain the Lindop committee made recommendations that a Data Protection Authority should be set up but the report has been gathering dust for over a year and its recommendations have yet to be implemented.

The considerations so far have concentrated on the commercial developments, but an important part of British society is concerned with manufacturing industry providing the wealth to support many of our other National activities. The effect of microelectronics on this sector of society will be no less significant and has been thus recognised by governments throughout the world. Most of the developed countries have introduced support schemes for the education of industry in the benefits of microelectronic technology and have provided direct funding for the initial developments. In Britain there is the Microprocessor Applications Project (MAP) which has three main functions: initial awareness, feasibility analysis and project support. Developments in the two aspects of products and processes for industry are eligible for support and many hundreds of projects at different levels have been supported. In the product sector, as already indicated, improved performance by using microelectronics is possible, without unduly increasing the price of the product. New products are also created, such as the calculator, television games, citizens' band radios, video tape records, etc. New and improved terminals such as the point-of-sale terminals and the salesman's remote terminal use the latest technology and are all increasing the ease with which people can have access to computers. The use of microelectronics is not restricted to classically electronic products as it has enormous impact, as yet

Electronic micrometer. The MICRO 2000 electronic micrometer provides the engineer with a fast and reliable method of making precision measurements over a range of 2.5 cm with an accuracy of 1 micron (one millionth of a meter). The measuring system is based on Moiré fringes, which are detected electronically and counted by the special chip. The number of fringes counted is directly proportional to the distance measured and the chip provides the necessary signals to illuminate the correct segments of the electronic display. *Photograph courtesy of Moore and Wright Ltd.*

only partially realised, on the mechanical industries. A good example of how microelectronics can transform an instrument is the electronic micrometer. The traditional micrometer, which is a small instrument used to make accurate measurements of thickness in engineering, is simply a refined screw-thread on which marks are cut to indicate the distance measured. It is however difficult to read and even with a skilled user, mistakes can easily be made. The benefits of the electronic version, which is now commercially available, is that more accurate results can be obtained and the measurement is displayed as a digital readout. This digital presentation of the measurement overcomes the problem of misreading and allows the instrument to be used by relatively unskilled people. The development of this instrument, however, highlights one of the main problems which traditional industry has in making use of microelectronics technology. The company that manufactures the instrument, one of Britain's leading micrometer makers, discovered that their expertise lay in the mechanical engineering field, while the electronic micrometer required skills in microelectronics and optics as well as mechanics. Initial design and development could be carried out by consultants but, in order to establish production, a whole section of the company's work force had to be re-educated or new labour employed to provide the new skills required. Although this is a particularly dramatic case of a purely mechanical company developing a relatively complex and high-technology product, it illustrates the general direction in which manufacturing industry must move if it is to produce competitive products. For this they require enlightened management, willing to accept and acknowledge their initial ignorance of the subject, enthusiastic to learn and to investigate the possibilities of exploiting microelectronics, plus an adequate supply of labour with the required electronic skills.

The second aspect of industry that is being dramatically affected is the control of manufacturing processes. The use of microelectronics, first in the form of sensors to measure various parameters of a process, followed by process calculations using microprocessors and finally control by distributed or centralised computers, allows higher quality goods to be produced more reliably and at a lower cost. Through the use of microelectronics much finer control can be exercised over the various stages of a process, thus reducing the wasteful use of expensive energy, detecting at an earlier stage of manufacture any potential failures and keeping a record of the progress during the manufacture of an item. The real-time analysis of information from a complex industrial plant such as an oil refinery can now be used to control the plant dynamically, so that changes in the conditions of manufacture can be accommodated without altering the final product.

Microelectronics also enhances dramatically the quality assurance that can be achieved as more automated test and inspection is carried out during manufacture. A consequence of the use of more sophisticated test equipment is that the skills required will change. For example, instead of making numerous individual mechanical measurements, the dimensions of a component could be continuously monitored during fabrication and the measurements analysed by a microprocessor system. The inspection engineer using such equipment would then be required to know how to program the machine to make the necessary measurements, and to analyse a more comprehensive range of results than would have been made by mechanical means.

This example highlights a growing problem, which was touched upon earlier, namely that the application of computers or microcomputer technology requires an investment in software development and the supply of suitably trained people falls far short of the market demand. It has been calculated that in the United States the demand for software engineers is such that if present trends continue, half the population will be required to write computer programs by the year 2000! This prediction, of course, assumes that only existing techniques of software development are available and this obviously will not be the case. However, it is still true that in most high-technology electronics companies and research laboratories, a large percentage of the effort is directed towards software development.

As far as Great Britain is concerned, there is a gross shortage of software engineers able to produce vitally important applications software. This country has traditionally been competent in software development and we are relatively well placed in comparison to many other developed countries in the supply of computer scientists. A knowledge of, and a capability in, computer programming is, however, going to be required much more widely than at present. It is easy to consider new labour with the correct skills being trained at colleges but consideration must also be given to those with existing skills that are no longer required. It is a simplistic view to expect all of these individuals to undergo retraining – although for some this will be possible – and it is clear that some redundancies must occur directly through the introduction of microelectronic technology. Equally clear, however, is the fact that without this appropriate use of microelectronics in products, such products will rapidly become uncompetitive with obvious results for the company involved. There is also a positive side to this picture, which is that through microelectronics many new products and processes are being developed, creating the next generation of jobs. For instance in the traditional industries, which may incorporate microelectronics to improve their manufacturing processes, jobs are created in maintenance and data analysis whereas the number of direct operators is usually reduced.

The major problem in the short term is producing the required number of electronic engineers and technicians to allow industry to develop along this path. Special government initiatives will help but the real solution must lie with the schools, which must become more receptive to the concepts of the technology in their teaching.

Returning to industry, widespread computerisation is already under way in many areas of industry and this will form the first stage of automation. The second stage removes the need for human intervention in many processes with the introduction of robots to carry out the repetitive tasks. The potential of robotics has been appreciated and research has been carried out for a number of years. Up until now, however, there has been little prospect of implementing serious robots because the power and size of the necessary computing elements has been too large. With the future promise of microelectronic technology this field is ripe for development. At present the robots used in industry are extremely limited in their functions, usually carrying out paint spraying or simple welding operations. More advanced systems can pick up an object and place it in a pre-determined position. However, no real intelligence is included to recognise or sort components and to take alternative action if, say, a bolt is presented upside down. In the near future, electronic eyes will be incorporated and the level of intelligence increased many times, with robots bristling with microprocessors at every joint. We are still a long way from 'Dr Who', although the possibility of a robotic dog such as 'K9' is within the capability of existing technology. As with the introduction of microprocessor-controlled equipment, the use of robots will have a dramatic effect on employment levels and cause a transfer of skills from more menial tasks to those requiring higher levels of training. At the highest levels of design, the process will become increasingly complex but will be aided by the use of computers. At the lower levels of manufacture and maintenance, automation using computers and later, robots, will be introduced, requiring more advanced knowledge. These trends would seem to differ little from the technological progress of society through the years.

With the growing automation of industry and hence the reduced requirement for human physical skills, the attention of humans is being directed towards the service industries. So far only the newly automated industrial sector, together with the information transactions industry can create wealth. However, there is scope for improving the available service activities, for instance those that are directly beneficial, such as medical and educational activities, and those that are purely recreational and cultural.

The rapid advance of microelectronics technology makes it increasingly difficult to predict the projected lifetime of existing jobs. We can anticipate that, over the next ten-year period, microelectronics will create significant and far reaching changes in our lifestyle, and our workstyle. For example, the growth in computerised typesetting in newspapers coupled with advances in television and telephone-based information systems such as Ceefax, Oracle and Prestel, and the development of video tape recorders will inevitably cause a shift in the labour profile of the newspaper industry even to the extent that newspapers as we know them today might be made obsolete with news punched into a local computer anywhere in the world being rapidly available on cheap, portable television sets receiving their signals from satellites.

The view that technological advances will create redundancies (on a personal/individual level) need not necessarily be associated with the view that such technological advances will create unemployment (on a national level). History has shown, through the agricultural and industrial revolutions that, although redundancies were created and employment profiles changed, from primary employment in fishing and agriculture to secondary employment in factories and engineering and presently to tertiary employment in the service industries and commerce, the overall national employment levels have re-equalised in the face of such redundancies. Thus although present high technology will inevitably create redundancies, it is to be anticipated that on a national scale these redundancies will be equalised by a new trend towards quaternary employment and the main quaternary industries can be anticipated as being education and leisure.

Returning to consideration of the newspaper industry we have already witnessed extreme labour-relations problems with regard to both computerised typesetting (and electronic newsgathering) – and ultimately redundancies will be created. However, the same technology that creates these redundancies is fostering new concepts in advertising methods in newspapers and on television – thus although redundancies will be created, the overall employment level will equalise. The problem that exists with such re-equalisation is that the individual workers made redundant may not possess the relevant 'up-market' skills to fill the newly created jobs. So the redundant workers will turn to a quaternary level of employment such as education, for retraining in these skills, or to leisure where they attempt to improve their quality of life. Indeed the day may come when a nation's wealth is measured by the composite knowledge of its population as opposed to its industrial input. Whether the present structure of society is able to accommodate the changes that are taking place, particularly at the rate with which they are occurring, is a question of fundamental importance to society and will form a topic for debate over many years to come.

What type of society then can we envisage as a result of microelectronic technology? Certainly not the total collapse predicted by some commentators. From a technical viewpoint major developments will be witnessed in the convergence of communications and computing to form the Information Society, in which the commodity 'information' will be traded in the way more tangible commodities are at present. Manufacturing industry will obviously continue to produce the products that society needs. Microelectronics can be brought to the aid of manufacture as previously discussed, and it will generate new products in its own right. The adoption of electronics by industry will produce a change in skill requirements, which provides a challenge both to management and to the education system. The important question, of course, is 'will the creation of new jobs balance the obsolescence in existing employment patterns?' Many studies have been carried out on this topic but no clear picture has emerged. Certainly, if a narrow view is taken, massive job loss can be predicted. On the other hand, in other areas, as we have seen with software, serious manpower shortages exist and, if the broad picture is considered, there is every reason to expect that the microelectronics revolution is but one stage in the technological programme that was started by the scientists of the nineteenth century and is likely to continue into the future.

FURTHER READING

Tom Forester (ed). *The Microelectronics Revolution.*
Basil Blackwell 1980. ISBN 0631 12572 8

Scientific American. *Microelectronics.*
W. H. Freeman 1977. ISBN 07167 0066 2

Iann Barron and Ray Curnow, *The Future with Microelectronics.*
Frances Pinter 1979. ISBN 089397 055 7

Christopher Evans. *The Mighty Micro.*
Gollancz 1979. ISBN 0575 02708 8

INDEX